Internet of Things

Technology, Communications and Computing

Series Editors

The series Internet of Things - Technologies, Communications and Computing publishes new developments and advances in the various areas of the different facets of the Internet of Things. The intent is to cover technology (smart devices, wireless sensors, systems), communications (networks and protocols) and computing (theory, middleware and applications) of the Internet of Things, as embedded in the fields of engineering, computer science, life sciences, as well as the methodologies behind them. The series contains monographs, lecture notes and edited volumes in the Internet of Things research and development area, spanning the areas of wireless sensor networks, autonomic networking, network protocol, agent-based computing, artificial intelligence, self organizing systems, multi-sensor data fusion, smart objects, and hybrid intelligent systems.

Internet of Things is covered by Scopus.

More information about this series at http://www.springer.com/series/11636

Maryam Farsi · Alireza Daneshkhah ·
Amin Hosseinian-Far · Hamid Jahankhani
Editors

Digital Twin Technologies and Smart Cities

Springer

Editors
Maryam Farsi
Through-life Engineering Services Institute
Cranfield University
Cranfield, Bedfordshire, UK

Amin Hosseinian-Far
Department of Business Systems
and Operations
University of Northampton
Northampton, Northamptonshire, UK

Alireza Daneshkhah
Faculty of Engineering, Environment
and Computing
Coventry University
Coventry, West Midlands, UK

Hamid Jahankhani
Northumbria University, London
London, Greater London, UK

ISSN 2199-1073 ISSN 2199-1081 (electronic)
Internet of Things
ISBN 978-3-030-18731-6 ISBN 978-3-030-18732-3 (eBook)
https://doi.org/10.1007/978-3-030-18732-3

This Springer imprint is published by the registered company Springer Nature Switzerland AG
The registered company address is: Gewerbestrasse 11, 6330 Cham, Switzerland

Foreword

Since its early days, the Internet has heralded a communications revolution that has evolved to a global backbone offering a huge potential for equitable connectivity for all from multinational enterprises to the individual entrepreneurs. This level of global reach at insignificant effort and cost has brought about a plethora of untapped opportunities alongside many threats and risks as evident from the cyberattacks and a burgeoning cybersecurity industry.

One key oversight over the use of emerging and advanced technologies is underestimating or neglecting the downside risks and potential harms that could arise for the adoption or implementation because of biased focus on the promises or upside benefits. The optimal position is naturally a balanced and equitable approach to identification and assessment of the so-called potential risks and rewards to ensure benefits are accrued whilst harms and risks are also taken into account and controlled at an acceptable level to the key stakeholders.

The other major oversight in the adoption of new innovations and technologies is the conventional focus on the price, cost, quality, reliability and performance at the expense of recognition of the social impact and potentially undermining human values and ethical norms. This is a more contextual concern for deployment of products, services and systems that is gaining prominence, especially in the web-based environments and services.

The advanced IoT, Smart Cities and Machine Learning, Digital Twins and Autonomous Systems applications are not immune to this precautionary paradigm. The promises of these innovations must be cultivated and explored alongside recognition of potential harms whether these may arise from the technology itself or its likely misuse. A balanced and human-centred approach to harnessing innovations whilst reducing risks of harm to the society and the environment should constitute a systematic and rational framework before the rush for rapid deployment.

After a few millennia, humanity has reached a stage that requires considerations beyond mere compliance with commercial, safety and security regulations. This is indeed the age of ethics and human-centric design.

London, UK
January 2019

Prof. A. G. Hessami

Preface

Origins and Use of the Book

Digital Twin (DT) technologies are already being used in different systems and industries such as manufacturing, construction, health care, aerospace, transportation, etc. Implementation of DT technologies is already underway, and it is believed that it will be growing rapidly in the upcoming decades. In September 2017, Gartner Inc. listed digital twins among the top 10 strategic technologies in 2017.

DT technologies enable us to create a virtual duplicate of our real system and therefore provide us with a platform to review activities, interactions and consequences of different decisions within the real system. In industry, DT technologies are being used to improve productivity, efficiency, availability and consequently quality of an asset or a service. By considering cities as a complex system, there will be numerous applications linked to the underlying technologies to develop digital twins. The inherent complexity of such applications and their interactions can also potentially bring numerous benefits. A smart city will have DT applications in many areas including but not limited to transportation, logistics, energy and power, communications and health care. The primary goal of DT application in smart cities will be to improve efficiency of systems and subsequently sustainability of such systems. For instance, with respect to the transportation and logistics perspective, the focus will be on improving the performance in terms of enhancing energy consumption and reducing harmful emissions, e.g., carbon footprint reduction. Moreover, utilising DT technologies, predicting and removing the bottlenecks between the different elements within a complex system would be the main focus regarding communication and healthcare perspectives. Implementing DT technologies and integrating them with the managerial processes would be key in improving policymaking and decision-making in a city. Internet of Things (IoT) will provide the connection between all the parties, organisations and elements in the smart city with DT technologies.

Considering a smart city as a system with so many interrelationships between different components, several challenges exist between the actual components and their DTs. These interactive challenges are mainly raised by the presence of many stochastic data with uncertainty. Moreover, there are many challenges related to digital infrastructure, implementation and societal acceptance, regulation and legislation which need to be tackled to enable the integration of DT technologies in smart cities. These complexities around the application of DT technologies in smart cities would raise lots of questions, discussions and solutions that we aim to address in this book.

London, UK

Maryam Farsi
Alireza Daneshkhah
Amin Hosseinian-Far
Hamid Jahankhani

Motivation

The motivation behind this book and its use are as follows:

1. To gather remarkable and fundamental concepts with regards to different applications of DT technologies in smart cities.
2. To discuss the DT applications in smart cities for asset management, and the use of Internet of things and artificial intelligence.
3. To pave the way for drawing a roadmap for implementation of DT technologies in smart cities with respect to the Triple Bottom Line.
4. To discuss and conclude the benefits, consequences and upcoming challenges of DT technologies implementation in smart cities.
5. To provide a guideline, experts recommendation and suggestions to address these challenges.
6. To demonstrate further research on DT technologies and smart cities to ensure safety and security and reliability.

Contents

Part I Digital Twins and Smart Cities

The Convergence of Digital Twin, IoT, and Machine Learning: Transforming Data into Action 3
Maninder Jeet Kaur, Ved P. Mishra and Piyush Maheshwari

A Novel Approach Toward Enhancing the Quality of Life in Smart Cities Using Clouds and IoT-Based Technologies 19
Kamta Nath Mishra and Chinmay Chakraborty

The Future of Mobility with Connected and Autonomous Vehicles in Smart Cities 37
Patrice Seuwou, Ebad Banissi and George Ubakanma

A Digital Twin Model for Enhancing Performance Measurement in Assembly Lines 53
Christos I. Papanagnou

Information Sharing in Sustainable Value Chain Network (SVCN)—The Perspective of Transportation in Cities 67
Luai Jraisat

Health Care in the Cyberspace: Medical Cyber-Physical System and Digital Twin Challenges 79
Jaime Ibarra Jimenez, Hamid Jahankhani and Stefan Kendzierskyj

Part II Internet of Things, the Digital Twin Enabler

Present Scenarios of IoT Projects with Security Aspects Focused 95
Aniruddha Bhattacharjya, Xiaofeng Zhong, Jing Wang and Xing Li

IoT Security, Privacy, Safety and Ethics 123
Hany F. Atlam and Gary B. Wills

CoAP—Application Layer Connection-Less Lightweight Protocol for the Internet of Things (IoT) and CoAP-IPSEC Security with DTLS Supporting CoAP 151
Aniruddha Bhattacharjya, Xiaofeng Zhong, Jing Wang and Xing Li

Some Computational Considerations for Kernel-Based Support Vector Machine 177
Mohsen Esmaeilbeigi, Alireza Daneshkhah and Omid Chatrabgoun

A Secure Hybrid RSA (SHRSA)-based Lightweight and Efficient Personal Messaging Communication Protocol 191
Aniruddha Bhattacharjya, Xiaofeng Zhong, Jing Wang and Xing Li

Contributors

Hany F. Atlam School of Electronics and Computer Science, University of Southampton, Southampton, UK;
Department of Computer Science and Engineering, Faculty of Electronic Engineering, Menoufia University, Menoufia, Egypt

Ebad Banissi Division of Computer Science and Informatics, School of Engineering, London South Bank University, London, UK

Aniruddha Bhattacharjya Department of Electronic Engineering, Beijing National Research Center for Information Science and Technology, Tsinghua University, Beijing, China

Chinmay Chakraborty Department of Electronics and Communication Engineering, Birla Institute of Technology, Ranchi, Jharkhand, India

Omid Chatrabgoun Department of Statistics, Faculty of Mathematical Sciences and Statistics, Malayer University, Malayer, Iran

Alireza Daneshkhah Faculty of Engineering, Environment and Computing, Coventry University, Coventry, UK

Mohsen Esmaeilbeigi Department of Statistics, Faculty of Mathematical Sciences and Statistics, Malayer University, Malayer, Iran

Hamid Jahankhani Northumbria University, London, London, Greater London, UK

Jaime Ibarra Jimenez Northumbria University, London, UK

Luai Jraisat Faculty of Business and Law, University of Northampton, Northampton, UK

Maninder Jeet Kaur Department of Engineering, Dubai International Academic City, Amity University Dubai, Dubai, UAE

Stefan Kendzierskyj Northumbria University, London, UK

Xing Li Department of Electronic Engineering, Beijing National Research Center for Information Science and Technology, Tsinghua University, Beijing, China

Piyush Maheshwari Department of Engineering, Dubai International Academic City, Amity University Dubai, Dubai, UAE

Kamta Nath Mishra Department of Computer Science and Engineering, Birla Institute of Technology, Ranchi, Jharkhand, India

Ved P. Mishra Department of Engineering, Dubai International Academic City, Amity University Dubai, Dubai, UAE

Christos I. Papanagnou Salford Business School, University of Salford, Manchester, UK

Patrice Seuwou Division of Computer Science and Informatics, School of Engineering, London South Bank University, London, UK

George Ubakanma Division of Computer Science and Informatics, School of Engineering, London South Bank University, London, UK

Gary B. Wills School of Electronics and Computer Science, University of Southampton, Southampton, UK

Xiaofeng Zhong Department of Electronic Engineering, Beijing National Research Center for Information Science and Technology, Tsinghua University, Beijing, China

Part I
Digital Twins and Smart Cities

The Convergence of Digital Twin, IoT, and Machine Learning: Transforming Data into Action

Maninder Jeet Kaur, Ved P. Mishra and Piyush Maheshwari

Abstract Digital twins, Internet of Things (IoT), block chains, and Artificial Intelligence (AI) may redefine our imagination and future vision of globalization. Digital Twin will likely affect most of the enterprises worldwide as it duplicates the physical model for remote monitoring, viewing, and controlling based on the digital format. It is actually the living model of the physical system which continuously adapts to operational changes based on the real-time data from various IoT sensors and devices and forecasts the future of the corresponding physical counterparts with the help of machine learning/artificial intelligence. We have investigated the architecture, applications, and challenges in the implementation of digital twin with IoT capabilities. Some of the major research areas like big data and cloud, data fusion, and security in digital twins have been explored. AI facilitates the development of new models and technology systems in the domain of intelligent manufacturing.

Keywords Digital twins · Internet of things (IoT) · Artificial intelligence (AI) · Machine learning · Big data · Cyber-physical systems (CPS)

1 Introduction

There had been various advancements in new generation information technologies like IoT, AI, big data, cloud computing, edge computing, etc. that have wide applications in smart manufacturing [1]. The advanced computing and analytics in the cyber world has opened a bright perspective to smart manufacturing. The increase in digitization of manufacturing opens up various opportunities. It is predicted by

M. J. Kaur (✉) · V. P. Mishra · P. Maheshwari
Department of Engineering, Dubai International
Academic City, Amity University Dubai, Dubai, UAE
e-mail: mkaur@amityuniversity.ae

V. P. Mishra
e-mail: vmishra@amityuniversity.ae

P. Maheshwari
e-mail: pmaheshwari@amityuniversity.ae

M. Farsi et al. (eds.), *Digital Twin Technologies and Smart Cities*,
Internet of Things, https://doi.org/10.1007/978-3-030-18732-3_1

Gartner in 2016 that more than 20 billion devices (majority from the manufacturing industry) would be connected to each other by 2020, which further will generate 40 zettabytes of data in raw form as unstructured, semi-structured, and unstructured [2, 3]. Hence, there is a need to organize, analyze, and extract information from this raw data to obtain valuable information with the use of advanced computing mechanisms and algorithms [4, 5].

In conventional approach, the designers use the computer-aided simulation and engineering tools to design and predict the life cycle and perform various physical testing mechanisms. They do optimize design to maximize performance and cut down design cost. But in this approach, there is a limitation on tolerances, strategies relationships amongst the configurations, planning, etc. [6]. However, the development of computing industry with artificial intelligence, faster processing, enhanced algorithms, and increasing computational power in the field of products and production line—digital twin—enable the ability of real-time control and digitization [7–10]. Physical object, process or system can be represented with the help of digital twin. With the combination of data and intelligence that represent the structure, context, and behavior of a physical system, it offers an interface that allows monitoring the past and present operation and makes prediction about the future [11]. Therefore, digital twin, integrate AI, software analytics, and machine learning data to create digital simulation models that update and change as their physical equivalents change. This provides real-time monitoring and updates from multiple sources at the same time. It creates virtual models for physical objects in the digital way to simulate their behavior [12]. The virtual models could understand the state of physical entities through sensing data, to estimate and analyze the dynamic changes. The digital twin would achieve the optimization of the whole production process [13].

This chapter is organized as follows. The concepts and architecture of digital twin is reviewed in Sect. 2. The applications and challenges are also discussed in this section. Section 3 discusses the related work in the area of digital twins. Smart and intelligent manufacturing with AI evolution is explored in Sect. 4, followed by conclusions.

2 Digital Twin—Concept and Architecture

The growth of advanced technologies is paving way for the smart cities, where all the physical objects will have embedded computing and communication capabilities so that they can sense the environment and communicate with each other to provide the services. These intelligent interconnections and interoperability are also termed as IoT or machine-to-machine (M2M) communications [14]. Some of the important domains of a smart city are the smart energy, smart home, smart transport system, and smart manufacturing. Because of the affordability and availability of the sensors and actuators, data acquisition has become relatively easier. Monitoring and diagnosing the manufacturing machines through the Internet is a challenging task. The convergence of the physical and virtual worlds of manufacturing is still one of the major challenges in the field of Cyber-Physical Systems (CPS), which needs more research. To tackle these challenges, Industry 4.0 was conceptualized [15], which mentioned that if the production systems are made intelligent and smart, they can function more efficiently [16, 17]. There have been many developments to enable this, one of which is digital twin [18].

"Digital twin" is a concept that creates a model of a physical asset for predictive maintenance. This model will continually adapt to changes in the environment or operation using real-time sensory data and can forecast the future of the corresponding physical assets [19]. It can monitor and identify potential issues with its real physical counterpart. In addition, it allows the prediction of the remaining useful life (RUL) of the physical twin by leveraging a combination of physics-based models and data-driven analytics. It consists of three main parts: (i) physical products in real space (ii) virtual products in virtual space, and (iii) the connections of data and information that will tie the virtual and real products together. Therefore, collecting and analyzing a large volume of manufacturing data to find the information and connections has become the key to smart manufacturing.

The concept of digital twin presented by Grieves at one of his presentations in 2003 on Product Lifecycle Management (PLM) at University of Michigan [20]. GE has started its digital transformation journey centered on Digital Twin, by building critical jet engine components that predict the business outcomes associated with the remaining life of those components [21].

The work done in [22] was the first initiative to come up with a dynamic Bayesian network approach for digital twin, where they utilized the concept of digital twin for tracking the evolution of time-dependent variables to monitor aircraft structure.

2.1 Architecture

The basic architecture of digital twin consists of the sensor and measurement technologies, Internet of Things, and machine learning. From the computational perspective, the key technology to propel a digital twin is the data and information fusion that facilitates the flow of information from raw sensory data to high-level understanding and insights [23]. The key functionality of digital twin implementation through physics-based models and data-driven analytics is to provide accurate operational pictures of the assets [24]. This helps the digital twin mirror the activities of its corresponding physical twin with the capabilities of early warning, anomaly detection, prediction, and optimization. The IoT system carries out real-time data acquisition through its smart gateway and edge computing devices. The preprocessed online sensory data is fused to feed the digital twin model. The offline data, after processing with text/data mining algorithms and then inputted to the digital twin as well. The offline computing resources utilized to train deep learning models. The digital twin combines modeling and analytics techniques to create a model of a specific target, e.g., flight critical component, etc. Hence, digital twin use is specified as predictive maintenance workflow to enable the delivery of accurate forecasting, using the data that is continuously acquired with IoT sensors via machine learning algorithms.

2.2 *Applications of Digital Twin*

Digital Twin determines the best course of action by eliminating the guesswork to service the critical assets in the manufacturing units. The increasing adoption of the IoT is ideal for enterprises to leverage digital twin platforms to boost their services and platforms. Some of the applications are given as follows [25]:

- Performance Optimization—Digital twin helps to determine the optimal set of parameters and actions that can help maximize some of the key performance metrics and provide forecasts for long-term planning. For example, NASA proposed and adopted for monitoring and optimization on safety and reliability optimizations of spacecraft [26, 27].
- Healthcare—Digital twin can be used for capturing and visualize a hospital system in order to create a safe environment and test the impact of potential changes on system performances. Not just operations, it also helps to improve the quality of health services delivered to the patients. For example, a surgeon can use it for a digital visualization of the heart, before opening it.
- Improve customer experience—As customers play a key role in influencing the strategies and decisions in any business. Enhancing the customer experience to retain and explore new customer base is the goal for the businesses. By directly creating a digital twin of the customer-facing applications, they can get feedback that boost the services directly offered to the customers.
- Maintenance—Digital twin can analyze performance data collected over time and under different conditions. For example, a racecar engine can be visualized to identify the required maintenance such as the component that is about to burn out.
- Machine Building—Digital twin is also used as a digital copy of the real machine that is created and developed simultaneously. Data from the real machine is loaded into the digital model to enable simulation and testing of ideas even before actual manufacturing starts.
- Smart Cities—Capturing the special and temporal implications to optimize urban sustainability. For example, "Virtual Singapore", a part of the Singapore government's smart nation Singapore initiative, is the world's first digital twin of an existing city-state, providing Singaporeans an effective way to engage in the digital economy and urbanization.

2.3 *Challenges of Digital Twin*

Some of the challenges to build and implement digital twins are as follows:

- The challenge to build a digital twin model combining product lifecycle management, manufacturing execution system and operations management system [28, 29]. After releasing process plans to the manufacturing execution system, using digital twin model in the cloud server to generate detailed work instructions associated with the production process design. Therefore, if there is any change from a production environment, the entire process is updated accordingly in the design and plan [30].
- Another challenge is how to build a more comprehensive digital-twin-driven physical-cyber-social connected production line [31–33]. The preliminary function of digital twin model is to help enterprises to design and manufacture of excellent products. However, the main aim of a digital twin model is to continue to accumulate knowledge of the design and manufacturing is reused and improved continuously [34].
- One of the most important challenges is to incorporate the big data analytics [35] into digital twin model. When directly collect real-time data from the production equipment, it will cover the information on the digital twin model. When compared to design with actual manufacturing result, the big data analytics are supposed to identify whether there is a difference and find out the cause of the differences [36, 37]. In addition, intelligent decoupling of combined problems is desirable.
- Currently, there are no optimized methods to integrate the different engineering models on the digital twin. There are data transfer mechanisms between domain-specific engineering tools. Besides technical reasons, a cross-domain collaboration also has a challenge of employing modularization methods as a multi-domain mechatronic system as viewed from a physically oriented or a function-oriented perspective [38].

3 Machine Learning, Artificial Intelligence, and IoT to Construct Digital Twins

Digital twin consists of the sensors and measurement technologies, IoT, simulation, and modeling and machine learning technologies. IoT devices are expected to generate a significant amount of data as their use becomes ubiquitous. IoT-cloud communication models and big data generated by devices results in increased latency and incremental data of cloud services and upstream data on behalf of IoT services.

3.1 Related Work

There are various fields, which contribute to the digital twin implementation—networking, cloud/edge computing, machine learning, sensors, etc. In the field of artificial intelligence, work done by [39] were the first one to initialize with a dynamic Bayesian network approach for digital twin, wherein they utilized the concept of digital twin for tracking the evolution of time-dependent variables to monitor aircraft structure.

In IoT world, AI will enhance the functionalities of digital twins in which a dynamic software model is formed of a physical thing or system that relies on sensor data to understand its state, respond to changes, improve operations, and add value. In [13], authors proposed digital-twin-driven product design, manufacturing, and service with big data, but their work has been mostly investigative in nature. Currently, the Industrial Internet or Industrial Internet of Things (IIoT) use digital twins for implementation in manufacturing industry. The work done in [40] discusses how IoT devices and IoT systems can be managed and optimized throughout their lifecycle using the mechanism of digital twins.

In manufacturing, IoT devices generate the data from product lifecycle, such as design, manufacturing, MRO, etc., [41]. Manufacturing data are generally from the following aspects [21]:

- Data from the manufacturing systems, e.g., MES, PDM, SCM, ERP, etc., and from other computer-aided systems like CAD/CAM, CAE, etc.
- Data from Internet/users, e.g., from e-commerce—Amazon, Walmart, Facebook, twitter, etc.
- Data from manufacturing equipment with respect to real-time performance, material of product data, environmental data, etc.

Processing of the collected data should go through various steps to extract the information. As the data collected via various ways like sensors, application-programming interface (API), software development kit (SDK), etc., undergoes cleaning before processing and analyzing [42–44]. This cleaned data integrates and stored for the exchange and sharing for manufacturing data at all levels. Further, the real-time data or offline data analysis and mining by advanced data analysis methods and tools like AI and machine learning, deep learning, etc. utilize cloud computing [45–47]. The valuable information extracted from large number of dynamic and fuzzy data enables manufacturers to deepen their understandings of various stages of product lifecycle. Hence, this helps the manufacturers to make more rational and informed decisions.

In Intelligent manufacturing (IM) area, the first book was published in 1988 [48], which resulted in the emergence of many methods, applications, and techniques in various areas of manufacturing like design, scheduling, production, control, modeling, testing, etc. [49]. In [49], the authors surveyed the relevant AI methods introduced in the field of manufacturing and grouped them as knowledge-based/expert systems, fuzzy logic, multi-agents, neural networks, evolutionary genetic algorithms, and simulated annealing. Introduction of knowledge-based/expert systems efficiently in computer integrated manufacturing (CIM) components but intelligent, manufacturing system (IMS) in industry were mainly in large companies [50]. The most famous IMS research was the international scheme of joint research called Intelligent Manufacturing System found in 1995 that influenced from dated back to 1989 from Japan [51]. In 90's, agent-based systems for intelligent manufacturing were developed followed by the web service-based systems for manufacturing and crowd-sourcing [52–55]. The agent-method seemed to be the potential solution as it offered a proper paradigm for the intelligent CIM components and IMS [56–58]. Intelligent agents are used in distributed AI and such an agent-based approach can handle the issues of the present software applications, specifically those working conditions that are highly dynamic and uncertain [59]. However, most agent-based systems are still at a research and prototype stage in labs and not widely adopted in manufacturing.

4 Intelligent and Smart Manufacturing with AI Evolution

Some of the key research areas, which we have studied in this chapter, are *Fusion of Big Data, Cloud and Cyber-Physical Systems, Information and Data Fusion in Decision-Making, Security in Digital Twins/Smart Manufacturing.*

4.1 Fusion of Big Data, Cloud and Cyber-Physical Systems

The cyber-physical systems (CPS) is another name for digital twin phenomena that makes possible the data analysis based control of the resources or physical environments with much ease. Here, the physical systems collect sensory information from the real world and send them to the digital twin computational modules through communication technologies (wireless). It is challenging to incorporate big data analytics into CPS [60]. The technologies used for the implementation of smart manufacturing span a wide spectrum of domains, which are initially referred to as the IoT technologies, and then many other related techniques such as the Internet of services (IoS), CPS, big data, and advanced robotics [61] have been a part. The rise of IoT/CPS and small objects (phones) has made the products more connected and accessible, from which the wealth of data generated allows accurate targeting and further enabling proactive management of enterprises through informed, timely in-depth decision execution [62]. Therefore, the fusion of human, data and smart and intelligent algorithms has far-reaching effects on manufacturing efficiency.

Collection, visualization, and analysis of the large volume of manufacturing data is the key to smart manufacturing. From the input of raw material to the output of finished products, the digital twin manages and optimizes the complete manufacturing process [63]. The virtual workshop or factory include the geometrical or physical models of operators, material, equipment, tools, environment, etc., as well as the behaviors, rules, dynamics, and many other factors [64]. The virtual model of product is created to establish the product digital twin. The product digital twin would always keep in company with the product to provide the value-added services [65]. Some of these are given as follows [66]:

- The product in use is monitored in real-time, as the product digital twin continuously records the product usage status, environmental data, operating parameters, etc.
- The virtual model can simulate the operation conditions of the product in different environments. Hence, it can confirm what effects the different environmental parameters and operation behaviors would have on health, lifetime, and performance to control the status and behavior of the physical product.
- Based on real-time data from the physical product and historical data, the product digital twin predicts the product remaining life, faults, etc.

Based on the predictions for health condition, remaining life and faults, the proactive maintenance is carried out to avoid the sudden downtime. In addition, when the faults occur with the high fidelity virtual model of the product, the fault would be visually diagnosed and analyzed, so that the position of the faulty part and the root cause of fault displayed to the users [67]. These operations–maintenance and repair operations (MRO) which include disassembly sequence, spare parts, etc. provide sustainability. Before starting the actual MRO, the walkthrough about MRO strategies executed in the virtual world based in the virtual reality and augmented reality to impose predictive analysis. As the virtual models faithfully reflect the mechanical structure of the parts and the coupling between each other, it can identify whether the MRO strategies are effective, executable and optimal. The data from the different stage of product lifecycle are accumulated and inherited to contribute to the innovation of the next generation product.

Moreover, in the design phase, product innovation relies on the accurate interpretation of market preferences and customer demands, in accordance with the optimal planning. Besides, once the design changes, the manufacturing process can be easily updated, including updating the bill of materials, processes, and assigning new resources. As a result, the convergence of digital twin, big data and service, enables the production, planning, optimizing and manufacturing process in real-time. In the daily operation and MRO of the product, the virtual models of physical products synchronize with the real state of the product through sensors. The operation status of the product and the health status of the components generated in real-time. In addition to the sensors data, digital twin also integrates the historical data, e.g., maintenance records, energy consumption, etc. and through the analysis of this data, product digital twin can continuously predict the state of the product and remaining life of the product and probability of faults. It can also analyze the unknown problems by comparing the actual product response and anticipating the product response in specific scenarios. Hence, it improves product life and maintenance efficiency and reduces the maintenance cost. Big data analytics is responsible for all the data acquired and analyzed by the smart manufacturing. Therefore, the convergence of the digital twin and the big data is very important for smart manufacturing [13].

4.2 Information and Data Fusion for Decision-Making

More generally, the definition of information fusion is "*the study of efficient methods for automatically or semiautomatically transforming information from different sources and different points in time into a representation that provides effective support for human or automated decision-making.*" [68]. Decision-making in big data is driven by predictions—learning from data (experience) to predict, and actions are taken in response to predictions [69]. Machine learning, which learns from data and uses statistical approaches to assist decision-making that operates well in practice, contrasts with the older expert system approach that aims to mimic the rules from human experts with the help of programmers translating the explicit rules into software code. Digital twin integrates the various data originating from the physical, cyber, and social spaces through information and data fusion techniques to provide human-understandable abstractions and inferences.

Data fusion with the multimodal data collected from heterogeneous data sources, advanced mining techniques may be necessary to fuse the data. The data collected may be in different scales of measurement [70]. This information and data fusion layer consists of various statistical or logic-based methods to integrate the outputs from the data processing layer to achieve a cohesive view of applications. The fusion techniques ensure that there is a combination of computers, smart devices, and people working together. Some of these techniques are given as follows [71]:

- Semantic Reasoning—Semantic web-based methods have been used to map proprietary relational datasets, environment monitoring data streams and participatory sensing data and this data then is combined (with match filters) with user preferences to form a dynamic social structure of things.
- Tensor Decomposition—The tensor-based methods exploit existing approaches for data fusion that can detect hidden information. This method is generally to analyze the behavior similarity of users. Group-centric data fusion is performed based on the approximate tensor, with each element in the approximate tensor representing the prevalence of the corresponding behavior in the group [72].
- Cross-space data fusion through correlation—Cross-space data fusion has taken the form of statistical methods, to calculate correlation between numerical data streams derived from the physical and social space. These include utilizing the data generated by citizens in social networking platforms in conjunction with data from sensor installations to build a model of the city's dynamics.

Decision support mechanisms consists of prediction algorithms that support further insights through data fusion. The flow of information from raw data to high-level decision-making propels by sensor-to-sensor, sensor-to-model, and model-to-model fusion. Therefore, manufacturers will make more rational, responsive, and informed decisions and enhance their competitiveness.

4.3 Security in Digital Twin

Bringing the Internet to the manufacturing industry offers opportunities but also new challenges. The required information flow across many communication networks raises questions about IT and data security that was not relevant when the machines were not programmable and were not connected to any other infrastructure except the power. Therefore, providing security or maintaining the security in the current manufacturing system in organizations are becoming a challenging task due to the cyberattacks and intrusions in current scenario. The security required for the manufacturing system for the following five levels as depicted in the CIM model [73]. CIM is a highly integrated model that has been used and incorporated into many models and standards in the manufacturing industry.

1. Enterprise/Corporate Level—At this level, the decisions related to operational management which define the work flows to produce the end product are made.
2. Plant Management Level—This level manages the decisions locally on the plant management network.
3. Supervisory Level—This level manages various manufacturing cells, each performing a different manufacturing process.
4. Cell Control Level—At this level, processes perform different actions.
5. Sensor Actuator Level—Here, the sensors, actuators, controllers integrate to perform the physical process.

Because of its design, this model is vulnerable to security attacks. The various protocols used to support this infrastructure—modbus, distributed network protocol (DNP3), industrial Ethernet, PROFIBUS, building automation and control networking (BACnet), etc. are only used for supervisory and control mechanisms but not security and lack mechanisms to provide authentication, integrity, freshness of the data, non-repudiation, confidentiality and measures to detect faults and abnormal behavior. Following are the cyber liabilities for most of the manufactures [74]: interruption in business, data breach, cyber extortion, intellectual property, third party damage.

Various solutions to counter these security discrepancies are:

- Public Key Infrastructure—To use device certificates and public key infrastructure (PKI) architectures. Implementing PKI into embedded systems secures the communication layer, creating a system that verifies the authenticity, configuration, and integrity of connected devices. This makes PKI ideal for large-scale security deployments that require a high level of security with minimal impact on performance [73].
- Encryption of the data—Highly confidential data must be encrypted to ensure that only authorized users have access by deploying anti-malware and hardening software on all IT and OT systems. In addition, use of symmetric encryption algorithms, hybrid encryption schemes, cryptographic hash functions, digital signatures, key agreement and distribution protocols are widely used to ensure only

authorized entities. The work done in [75, 76] proposes key management systems are studied and discussed.

- Intrusion detection systems—It is always necessary to monitor the dynamic behavior of the security systems and seek to find if there is an abnormal activity. Intrusion detection system (IDS) approaches tackle these issues. IDS are classified by source of data (audit source)—also called *network based* or *host based* and detection technique (the data needed for analysis)—also called *knowledge-based* or *behavior based*. Receiver operating characteristics (ROC) curve, which depicts the detection probability versus false alarm probability, evaluates the performance of IDS. Studies [77–79] show that most work in this area has been in behavior-based network intrusion systems since knowledge-based systems require detailed knowledge of previous exploits to define characteristics of the attack. Hence, IDS research for smart manufacturing and IoT systems is still in progress and face a lot of challenges due to limited testbed availability and insufficient data from real incidents.
- Policies and Regulations—There are various special guidelines to enforce security mechanisms in smart manufacturing systems [74]. Some of these guides are Guide to Industrial Control Systems (ICS) for SCADA systems, The National Institute of Standards and Technology (NIST), Distributed Control Systems (DCS), Department of Homeland Security (DHS), The Centre for the Protection of National Infrastructure (CPNI), etc.

Planning for security involves understanding the nature of threats, identifying vulnerabilities, quantifying the value to be lost if in case security breach happens and investing in security appropriately. This gives an autonomous model in which products and machines will become active participants in IoT behaving as autonomous agents throughout the production line.

5 Conclusion

Digital Twin has been recognized by many developed companies like GE, IBM, and Cisco as next-generation core infrastructure and are focusing more on developing CPS-related technologies and utilization of platforms. IoT and Artificial Intelligence in smart manufacturing was the initial step to recognize the sensors prerequisite into the machine parts from where the real-time analytics will get the data. Fusion of human, data and smart/intelligent algorithms has far-reaching effects on manufacturing efficiency. However, the intensive communication and high amounts of data involved also bring in new challenges. In this chapter, we discussed the architecture of the CPS, applications, and challenges involved in the implementation of Digital Twins. It also discusses the related work in the area of machine learning, artificial intelligence in the field of smart manufacturing. Furthermore, the key research areas—Fusion of Big Data, Cloud and Cyber-physical systems, Information and Data

Fusion for decision-making and Security in Digital Twins/Smart manufacturing were discussed.

We have discussed the connection between the data about the physical product and the information contained in the virtual product and its synchronization. By merging the virtual product information as to how the product manufacturing takes place, we can have an instantaneous and simultaneous perspective on how the manufactured product is meeting its design specification goals. Hence, information on product manufacturing is predicted and working in real-time as well is monitored. From the security point of view, the potential consequences of security attacks on smart manufacturing systems like, injuries, death, and damage to physical infrastructure, equipment, and the environment are likely to occur simply because the actuators in manufacturing system have connection to such things. It is important that the adoption of IoT and machine learning embed security from the start, integrated with functionality in smart manufacturing systems. Therefore, the convergence of IoT and Machine Learning with Digital twins will improve productivity, uniformity, and quality of the products.

References

1. Tao, F. and Qi, Q.: New IT driven service-oriented smart manufacturing: framework and characteristics. IEEE Trans. Syst., Man, Cybern. Syst. (2017)
2. Mourtzis, D., Vlachou, E., Milas, N.: Industrial Big Data as a result of IoT adoption in manufacturing. Procedia CIRP. **55**, 290–295 (2016)
3. Gantz, J. and Reinsel, D.: The digital universe in 2020: Big data, bigger digital shadows, and biggest growth in the Far East. IDC iView: IDC Analyze the future. 1–16 (2012)
4. Hashem, I.A.T., Yaqoob, I., Anuar, N.B., Mokhtar, S., Gani, A., Khan, S.U.: The rise of big data on cloud computing: Review and open research issues. Inf. Syst. **47**, 98–115 (2015)
5. Yi, S., Li, C. Li, Q.: A survey of fog computing: concepts, applications and issues. In: Proceedings of the 2015 workshop on mobile big data, ACM, pp. 37–42 (2015)
6. Leng, J., Jiang, P.: Dynamic scheduling in RFID-driven discrete manufacturing system by using multi-layer network metrics as heuristic information. J. Intell. Manuf. 1–16 (2017)
7. Wang, S., Wan, J., Zhang, D., Li, D., Zhang, C.: Towards smart factory for industry 4.0: a self-organized multi-agent system with big data-based feedback and coordination. Comput. Netw. **101**, 158–168 (2016)
8. Wang, S., Wan, J., Li, D., Zhang, C.: Implementing smart factory of industry 4.0: an outlook. Int. J. Distrib. Sens. Netw. **12**(1), 3159805 (2016)
9. Xu, Y., Sun, Y., Wan, J., Liu, X., Song, Z.: Industrial big data for fault diagnosis: Taxonomy, review, and applications. IEEE Access. **5**, 17368–17380 (2017)
10. Wan, J., Tang, S., Li, D., Wang, S., Liu, C., Abbas, H., Vasilakos, A.V.: A manufacturing big data solution for active preventive maintenance. IEEE Trans. Industr. Inf. **13**(4), 2039–2047 (2017)
11. Grieves, M.: Digital twin: manufacturing excellence through virtual factory replication. White paper. (2014)
12. Constante, T.A.D.S.L.: Contribution for a Simulation Framework for Designing and Evaluating Manufacturing Systems. (2018)
13. Tao, F., Cheng, J., Qi, Q., Zhang, M., Zhang, H., Sui, F.: Digital twin-driven product design, manufacturing and service with big data. Int. J. Adv. Manuf. Technol. **94**(9–12), 3563–3576 (2018)

14. Machine-to-Machine Communications (M2M): Impact of smart city activity on IoT environment, european telecommunications standards institute (ETSI). Sophia Antipolis, France (2015)
15. Schuh, G., Anderl, R., Gausemeier, J., Hompel, M.T., Wahlster W.: Industrie 4.0 maturity index. Managing the digital transformation of companies. Munich: Herbert Utz (2017)
16. Ribeiro, L., Björkman, M.: Transitioning from standard automation solutions to cyber-physical production systems: an assessment of critical conceptual and technical challenges. IEEE Syst. J. 1–13 (2017)
17. Qin, J., Liu, Y., Grosvenor, R.: A categorical framework of manufacturing for industry 4.0 and beyond. Procedia Cirp **52**:173–178 (2016)
18. Uhlemann, T.H.J., Lehmann, C., Steinhilper, R.: The digital twin: realizing the cyber-physical production system for industry 4.0. Procedia Cirp. **61**:335–340 (2017)
19. Tuegel, E.J., Ingraffea, A.R., Eason, T.G., Spottswood, S.M.: Int. J. Aerosp. Eng. **154798**:14 (2011)
20. Grieves, M.: Digital twin: manufacturing excellence through virtual factory replication. White Pap. (2014)
21. Qi, Q., Tao, F., Zuo, Y., Zhao, D.: Digital twin service towards smart manufacturing. Procedia CIRP **72**(1), 237–242 (2018)
22. Li, C., Mahadevan, S., Ling, Y., Wang, L., Choze, S.: A dynamic Bayesian network approach for digital twin. In: 19th AIAA Non-Deterministic Approaches Conference, p. 1566 (2017)
23. Liu, Z., Meyendorf, N., Mrad, N.: The role of data fusion in predictive maintenance using digital twin. In: AIP Conference Proceedings. **1949**(1):020023 (2018). AIP Publishing
24. Schmidt, M.T.: ANSYS Advant. XI:43–45, 2017
25. https://www.gavstech.com/wp-content/uploads/2017/10/Digital_Twin_Concept.pdf. Last accessed 21 Oct 2018
26. Grieves, M.: Digital twin: manufacturing excellence through virtual factory replication. White Pap. (2014)
27. Glaessgen, E., Stargel, D.: The digital twin paradigm for future NASA and US Air Force vehicles. In: 53rd AIAA/ASME/ASCE/AHS/ASC Structures, Structural Dynamics and Materials Conference 20th AIAA/ASME/AHS Adaptive Structures Conference 14th AIAA 1818 (2012)
28. Lee, J., Ardakani, H.D., Yang, S., Bagheri, B.: Industrial big data analytics and cyber-physical systems for future maintenance & service innovation. Procedia CIRP **38**, 3–7 (2015)
29. Lee, J., Bagheri,B., Kao, H.A.: A cyber-physical systems architecture for industry 4.0-based manufacturing system. Manuf. Lett. **3**:18–23 (2015)
30. Wang, X.V., Wang, L.: A cloud-based production system for information and service integration: an internet of things case study on waste electronics. Enterp. Inf. Syst. **11**(7), 952–968 (2017)
31. Barnaghi, P., Sheth, A., Singh, V., Hauswirth, M.: Physical-cyber-social computing: looking back: looking forward. IEEE Internet Comput. **3**, 7–11 (2015)
32. Hussein, D., Park, S., Han, S.N., Crespi, N.: Dynamic social structure of things: a contextual approach in CPSS. IEEE Internet Comput. **19**(3), 12–20 (2015)
33. Uhlemann, T.H.J., Lehmann, C., Steinhilper, R.: The digital twin: realizing the cyber-physical production system for industry 4.0. Procedia Cirp. **61**:335–340 (2017)
34. Schleich, B., Anwer, N., Mathieu, L., Wartzack, S.: Shaping the digital twin for design and production engineering. CIRP Ann. **66**(1), 141–144 (2017)
35. Lynch, C.: Big data: How do your data grow? Nature **455**(7209):28 (2008)
36. Bandaru, S., Ng, A.H., Deb, K.: Data mining methods for knowledge discovery in multi-objective optimization: Part A-Survey. Expert Syst. Appl. **70**, 139–159 (2017)
37. Bandaru, S., Ng, A.H., Deb, K.: Data mining methods for knowledge discovery in multi-objective optimization: Part B-New developments and applications. Expert Syst. Appl. **70**, 119–138 (2017)
38. Feldmann, S., Vogel-Heuser, B.: Änderungsszenarien in der Automatisierungstechnik–Herausforderungen und interdisziplinäre Auswirkungen. Engineering von der Anforderung bis zum Betrieb **3**:95 (2013)

39. Li, C., Mahadevan, S., Ling, Y., Wang, L., Choze, S.: A dynamic Bayesian network approach for digital twin. In: 19th AIAA Non-Deterministic Approaches Conference, p. 1566 (2017)
40. Canedo, A.: Industrial IoT lifecycle via digital twins. In: Proceedings of the Eleventh IEEE/ACM/IFIP International Conference on Hardware/Software Codesign and System Synthesis, pp. 29 (2016), ACM
41. Li, J., Tao, F., Cheng, Y., Zhao, L.: Big data in product lifecycle management. Int. J. Adv. Manuf. Technol. **81**(1–4), 667–684 (2015)
42. Zhang, Y., Zhang, G., Wang, J., Sun, S., Si, S., Yang, T.: Realtime information capturing and integration framework of the Internet of manufacturing things. Int. J. Comput. Integr. Manuf. **28**(8), 811–822 (2015)
43. Kumar, A., Kim, H., Hancke, G.P.: Environmental monitoring systems: a review. IEEE Sens. J. **13**(4), 1329–1339 (2013)
44. Wang, S., Li, Y., Liu, N., Wang, S.: Noisy-data-disposing algorithm of data clean on the attribute level. Comput. Eng. **9**, 031 (2005)
45. Zhu, C., Wang, H., Liu, X., Shu, L., Yang, L.T., Leung, V.C.: A novel sensory data processing framework to integrate sensor networks with mobile cloud. IEEE Syst. J. **10**(3), 1125–1136 (2016)
46. Lin, K.W., Deng, D.J.: A novel parallel algorithm for frequent pattern mining with privacy preserved in cloud computing environments. Int. J. Ad Hoc Ubiquitous Comput. **6**(4), 205–215 (2010)
47. Siddiqa, A., Hashem, I.A.T., Yaqoob, I., Marjani, M., Shamshirband, S., Gani, A., Nasaruddin, F.: A survey of big data management: taxonomy and state-of-the-art. J. Netw. Comput. Appl. **71**, 151–166 (2016)
48. Wright, P., Bourne, D.A.: Manufacturing Intelligence. Addison-Wesley, Boston, MA (1988)
49. Teti, R., Kumara, S.R.T.: Intelligent computing methods for manufacturing systems. CIRP Ann. **46**(2), 629–652 (1997)
50. Kopacek, P.: Intelligent manufacturing: present state and future trends. J. Intell. Rob. Syst. **26**(3–4), 217–229 (1999)
51. Setoya, H.: History and review of the IMS (Intelligent Manufacturing System). In: 2011 International Conference on Mechatronics and Automation (ICMA), pp. 30–33 (2011)
52. Shen, W., Norrie, D.H.: Agent-based systems for intelligent manufacturing: a state-of-the-art survey. Knowl. Inf. Syst. **1**(2):129–156 (1999)
53. Mostafaeipour, A., Roy, N.: Implementation of web based technique into the intelligent manufacturing system. Int. J. Comput. Appl. **17**(6), 38–43 (2011)
54. McAfee, A.P.: Enterprise 2.0: the dawn of emergent collaboration. MIT Sloan Manag. Rev. **47**(3):21 (2006)
55. Estellés-Arolas, E., González-Ladrón-De-Guevara, F.: Towards an integrated crowdsourcing definition. J. Inf. Sci. **38**(2), 189–200 (2012)
56. Madejski, J.: Survey of the agent-based approach to intelligent manufacturing. J. Achiev. Mater. Manuf. Eng. **21**(1), 67–70 (2007)
57. Monostori, L., Váncza, J., Kumara, S.R.: Agent-based systems for manufacturing. CIRP Ann. Manuf. Technol. **55**(2), 697–720 (2006)
58. Leitão, P.: Agent-based distributed manufacturing control: a state-of-the-art survey. Eng. Appl. Artif. Intell. **22**(7), 979–991 (2009)
59. Abbas, H.A., Shaheen, S.I., Amin, M.H.: Simple, flexible, and interoperable SCADA system based on agent technology (2015). arXiv preprint arXiv:1509.03214
60. Lynch, C.: Big data: How do your data grow? Nature **455**(7209), 28 (2008)
61. Louchez, A., Wang, B.: From Smart Manufacturing to Manufacturing Smart (2014)
62. Leiva, C.: On the Journey to a Smart Manufacturing Revolution. Ind. Week (2015)
63. Rosen, R., Von Wichert, G., Lo, G., Bettenhausen, K.D.: About the importance of autonomy and digital twins for the future of manufacturing. IFAC-PapersOnLine **48**(3), 567–572 (2015)
64. Tao, F., Cheng, Y., Cheng, J., Zhang, M., Xu, W., Qi, Q.: Theories and technologies for cyber-physical fusion in digital twin shop-floor. Comput. Integr. Manuf. Syst. (2017)

65. Zhuang, C., Liu, J., Xiong, H., Ding, X., Liu, S., Weng, G.: Connotation, architecture and trends of product digital twin. Comput. Integr. Manuf. Syst. **23**(4), 753–768 (2017)
66. Tuegel, E.J., Ingraffea, A.R., Eason, T.G., Spottswood, S.M.: Reengineering aircraft structural life prediction using a digital twin. Int. J. Aerosp. Eng. (2011)
67. Gockel, B., Tudor, A., Brandyberry, M., Penmetsa, R., Tuegel, E.: Challenges with structural life forecasting using realistic mission profiles. In: 53rd AIAA/ASME/ASCE/AHS/ASC Structures, Structural Dynamics and Materials Conference 20th AIAA/ASME/AHS Adaptive Structures Conference 14th AIAA, p. 1813 (2012)
68. Liggins II, M., Hall, D. and Llinas, J. eds.: Handbook of Multisensor Data Fusion: Theory and Practice. CRC press (2017)
69. Boström, H., Andler, S.F., Brohede, M., Johansson, R., Karlsson, A., Van Laere, J., Niklasson, L., Nilsson, M., Persson, A., Ziemke, T.: On the definition of information fusion as a field of research (2007)
70. Stevens, S.S.: On the theory of scales of measurement. Science **1946**(103), 677–680 (1946)
71. De, S., Zhou, Y., Larizgoitia Abad, I., Moessner, K.: Cyber–physical–social frameworks for urban big data systems: a survey. Appl. Sci. **7**(10):1017 (2017)
72. Zhang, Y.: GroRec: a group-centric intelligent recommender system integrating social, mobile and big data technologies. IEEE Trans. Serv. Comput. **9**(5), 786–795 (2016)
73. Tuptuk, N., Hailes, S.: Security of smart manufacturing systems. J. Manuf. Syst. **47**, 93–106 (2018)
74. https://industrytoday.com/article/5-types-of-cyber-liabilities-for-manufacturers/ Last accessed 24 Sep 2018)
75. Piètre-Cambacédès, L., Sitbon, P.: Cryptographic key management for SCADA systems-issues and perspectives. In: 2008 International Conference on Information Security and Assurance, pp. 156–161 (2008). IEEE
76. Pal, O., Saiwan, S., Jain, P., Saquib, Z., Patel, D.: Cryptographic key management for SCADA system: An architectural framework. In: 2009 International Conference on Advances in Computing, Control, & Telecommunication Technologies, pp. 169–174 (2009). IEEE
77. Roosta, T., Nilsson, D.K., Lindqvist, U., Valdes, A.: An intrusion detection system for wireless process control systems. In: 5th IEEE International Conference on Mobile Ad Hoc and Sensor Systems, pp. 866–872 (2008). IEEE
78. Carcano, A., Coletta, A., Guglielmi, M., Masera, M., Fovino, I.N., Trombetta, A.: A multidimensional critical state analysis for detecting intrusions in SCADA systems. IEEE Trans. Industr. Inf. **7**(2), 179–186 (2011)
79. Shin, S., Kwon, T., Jo, G.Y., Park, Y., Rhy, H.: An experimental study of hierarchical intrusion detection for wireless industrial sensor networks. IEEE Trans. Industr. Inf. **6**(4), 744–757 (2010)

A Novel Approach Toward Enhancing the Quality of Life in Smart Cities Using Clouds and IoT-Based Technologies

Kamta Nath Mishra and Chinmay Chakraborty

Abstract The smart city means using information technologies as per the needs of citizens in order to improve their day-to-day activities with high efficiency and decrease the living cost. The development of the smart city is the process of urbanization which can further improve the efficiency, reliability, and security of a city. The integration of communication and information technologies with the Internet of Things (IoT) and artificial intelligence (AI) techniques will be helpful for the urban/metro city areas in the overall management of schools, colleges, universities, libraries, power plants, transportation systems, waste management, hospitals, water supply, law enforcement, and other community services. The information and digital technologies will be used by end users and office administrations for the overall management of the things related to urban/metro city areas. The information and communication technologies (ICT) will allow officials of the city to interact/communicate directly with social communities and the infrastructure of the city will be available to the city officials on their fingertips. This chapter describes the economic benefits, implementation costs, and challenges toward the development of a smart city and its integration with cloud computing, IoT, and AI technologies. In this research work, we have tried to study the existing technologies, and we have proposed a novel architecture of a smart city which incorporates IoT, AI, and distributed cloud computing technologies and the smart city will have its own independent self-management system for managing almost everything related to the needs of our daily life. The proposed work will be helpful in maintaining the ecological system of the earth and the use of clean solar energy is making it friendly to the environment.

K. N. Mishra
Department of Computer Science and Engineering, Birla Institute of Technology, Ranchi 814142, Jharkhand, India
e-mail: mishrakn@yahoo.com

C. Chakraborty (✉)
Department of Electronics and Communication Engineering, Birla Institute of Technology, Ranchi 814142, Jharkhand, India
e-mail: cchakrabarty@bitmesra.ac.in

M. Farsi et al. (eds.), *Digital Twin Technologies and Smart Cities*, Internet of Things, https://doi.org/10.1007/978-3-030-18732-3_2

Keywords Cloud computing · E-services · Internet of Things · Internet/web technologies · Smart city architectures · Smart services · Urbanization · Urban technologies

1 Introduction

The integration of information and communication technology with the Internet of Things (IoT) will be helpful for the urban/metro city areas in the overall management of schools, colleges, universities, libraries, power plants, transportation systems, waste management, hospitals, water supply, law enforcement, and other community services. The information and digital technologies will be used by end users and office administrations for the overall management of the things related to urban/metro city areas. The ICT will permit city officials to interact/communicate directly with social communities and the infrastructure of the city will be available to the city officials on their fingertips. Further, different events and happenings of the city will be monitored with the help of ICT and IoT. The data and information will be collected from citizens by using the sensors integrated with real-time monitoring systems. The analysis of collected data will be helpful in maintaining the law and order situation of the city [1, 2].

The ICT with IoT will be used to improve the quality of life, quality services, traffic management, performance management, and interactivity between different service providers of smart cities. The ICT will be used to reduce costs and measure resource consumption. Further, the ICT will improve the contact between citizens and government and therefore, the deadlock which brings office strikes and agitations will be prevented [3]. The smart city applications will be developed to manage and control the flow of communication and daily life-related things in the real-time environment and it allows real-time responses between different components of the system [4].

In these days, some implementations of modern ubiquitous sensor network frameworks have been demonstrated for the actual installation of the smart metering facility and environmental protection and monitoring systems. The continuing activities are being extended to its border of machine-to-machine scenarios. The implementation and installation of ubiquitous sensor network platforms have shown a dynamic potential to establish a group of new services including internetworking and IoT. With reference to the IoT, the ubiquitous sensor network framework is currently being linked with the addition of new capabilities, and its integration with other components [5].

The cardinality of numerous stakeholders actively associated in the smart city market is so huge that many nontechnical and technical factors must be taken into considerations like end users, public administrations, and vendors, etc. In this respect, it is not clear that how international politics, business, and technology needs will be clubbed together to achieve the goal. In these days, no one is trying in this direction to achieve the goal [6].

This chapter describes the economic benefits, implementation costs and challenges toward the development of a smart city and its integration with cloud computing, IoT and artificial intelligence (AI) technologies. In this research work, the authors have presented the study of existing smart city technologies. In this chapter, the authors have proposed a novel architecture of a smart city which incorporates IoT, AI, and distributed cloud computing technologies. The proposed architecture of the smart city will have its own independent self-management system for managing almost everything related to the needs of our daily life. The proposed work will be helpful in maintaining the ecological system of the earth and the use of clean solar energy is making it friendly to the environment.

The whole chapter is organized into six sections, namely introduction (Sect. 1), literature review (Sect. 2), proposed cloud-based framework (Sect. 3), proposed the architecture for improving the quality of life in Smart cities (Sect. 4), discussions (Sect. 5), and the concluding remarks presented in Sect. 6.

2 Literature Review

In the last 15 years, the concept of the smart city has become very much popular in almost all parts of this earth and the people have started thinking about innovative ways of developing smart cities. The papers, articles, and reports on this topic have been exponentially increasing in the twenty-first century. The trend regarding publishing research articles related to smart cities and its indexing in standard databases like Scopus and DBLP has become a fashion in a current era [1, 2, 5, 6]. The identification of the core elements in smart city development is very important for researchers to understand that how different components of an urban/metro city areas offer unconventional electronic data interchange (EDI)-based e-services and communications. The smart city concept was introduced at the time when the whole world was struggling for coming out of worst economic crisis and the countries needed the help and cooperation of each other for the purpose of their survival. In the year 2008, IBM initiated smarter planet concept in which the smart interaction between different components of smart cities was proposed and by the beginning of 2009, the concept of the smart planet became viral in the whole world [7–9].

The initiative of being smart city was developed by European countries, as their cities tend to be densely populated and have better public transit. They are largely committed to their health and therefore, they have a very special space for cycling and walking and walking in their cities. They have a stronger focus on sustainability and low carbon emissions. The climate is changing drastically due to global warming. Therefore, critical thinking on developing environment-friendly communication system is required in the current age of new technology developments. The rapid increase in population and resource exhaustion are another area of deep thinking. Adverse effects of increasing urbanization is also a reason to ponder upon [10, 11].

Now, with the help of the internet, it is becoming possible for everyone to study whenever and wherever he/she want, and it does not require the viewer's presence at any particular time. Hence, the IoT helps both the consumer and the producer. Today, there is a requirement for a long-term method of developing sustainable cities by managing the life cycles of cities through improving economic performance over the entire life cycle. It provides opportunities by introducing healthy competition in terms of online services like waste management, education, healthcare, safety, and transportation systems, etc. [12, 13].

Further, the next-generation Internet potential with the help of IoT and information as a service (IoS) for generating forthcoming actual life applications and services is very vast in the context of smart city projects. The initial success of IoT deployments in smart city applications is jeopardized because of the unavailability of test beds of the desired scale and its suitability for the validation of most recent research outcomes. Many of the accessible test beds just offer limited testing environment up to a small domain of specific cases of deployments [14, 15].

3 Proposed Cloud-Based Approach for Enhancing the Life Quality of People in Smart Cities

The practice of using remote servers-based internetworks to store, manage, and process data rather than using a local server or external disks is known as cloud computing. In today's era of digitization and virtualization, cloud computing is an emerging trend to maintain and deploy software. World fame companies and industries such as Google, IBM, Microsoft, and Amazon have switched on to the cloud computing concept of managing and processing data. The concept of virtualization has brought a revolution in the existing technology. With virtual operating systems, the concept of virtual storage has been adopted so fast in the past few years that now their existence does not seem anything unnatural. To some extent, the virtual world has taken over the real world, and so is the concept of cloud computing. One exciting feature of cloud computing is that the customers of these services do not possess the resources but pay for them on a per-user basis. The services are provided to the customers on demand as a service via the internet [16].

The components of cloud computing consist of three different layers of services, namely infrastructure as a service (IaaS), platform as a service (PaaS), and software as a service (SaaS). The SaaS is a software sharing model in which applications are hosted by a service provider or a vendor, and the necessary services are provided by vendors to the customers through a wired/wireless network. It is usually implemented to make available business software functionality to venture the customers at very low cost while permitting the customers to get the same profit without thinking about the linked complexity of software/hardware installation, licensing, and support management [17, 18]. The SaaS vendor can keep the software applications on his/her own confidential server or install it on a cloud computing infrastructure managed/maintained by others (Amazon, Google, etc.). The integration of cloud computing with the concept "pay-as-much-as-you-use" method provides the application service provider to condense the investment in infrastructure services, and it enables the IoT-based cloud computing system to focus on providing further better services to the clients. In addition to it, there are also security issues with the SaaS model of cloud computing which needs to be recognized, identified and fixed before they are used as a service [19].

The PaaS is a compilation of linked services for creating and deploying software on cloud computing systems. Hence, it is not a singleton set-based technology. The PaaS has the capability to manage/control user subscriptions, security, resource metering, and distribution of other services. The PaaS plays a key role in cloud computing systems because it brings custom software development to the clouds. The **NIST** defined PaaS as "The capability provided to the consumer to deploy onto the cloud infrastructure consumer-created or acquired applications created using programming languages and tools supported by the provider" [20]. Some of the well-known PaaS-based cloud systems are Google App Engine (GAE), Microsoft Windows Azure (MWA), and Ground Operating Systems (GOS) [21].

The IaaS Internet protocols administer a huge set of computing resources including processing and storing capabilities. The IaaS is able to split, assign, and dynamically resize the available resources to develop ad hoc systems as per the requirements of clients. This is the exact scenario of IaaS. Further, in the cloud computing environment, the PaaS customers can be considered as application developers who are responsible for the design, development, and implementation of application software products in a cloud-based environment. The application developers who are responsible for uploading applications into the cloud and configure, monitor and manage deployed applications into the cloud are also the part of PaaS. Hence, the PaaS patrons can be paid according to the number of PaaS users; the number of processing tasks, storage capacity, duration of platform usage, and amount of network resources consumed by PaaS applications [22, 23].

Some of the severe disadvantages of PaaS are data security, limited flexibility, customer's captivity (a vendor lock-in time is usually the norm which can border the client's choices), and integration problems. The merger of PaaS services with different types of cloud and other applications may cause an unexpected and uncontrolled increase in the complexity of cloud computing systems [24, 25]. Hence, the researchers of the cloud community need to think about developing and maintaining an errorless cloud computing-based communication system in the forthcoming years.

4 Proposed Architecture for Quality of Life Improvement in Smart City

With the invention of sensor networks and artificial intelligence-based modern technologies, the future of smart city systems is predicted very bright. Therefore, the people in every part of the world are moving from rural to urban city areas and from urban areas to smart cities. The author has proposed smart city architecture, as shown in Fig. 1 where cloud computing technologies and sensor networks are integrated with artificial intelligence technologies to achieve the goals of high-level customized services in the real-time eco-friendly environment [26, 27].

As presented in Fig. 1, we need future components of smart city framework by having smart education, smart government, smart transportation, smart parking, smart healthcare, smart energy, eco-friendly environment, smart security, smart office, smart residential buildings, smart industries, and smart administration [28, 29]. For achieving these goals, the author has proposed to deploy the sensor nodes in each smart city domain because the sensor nodes will provide the primary data source for generating heterogeneous information. The information originated via sensor nodes

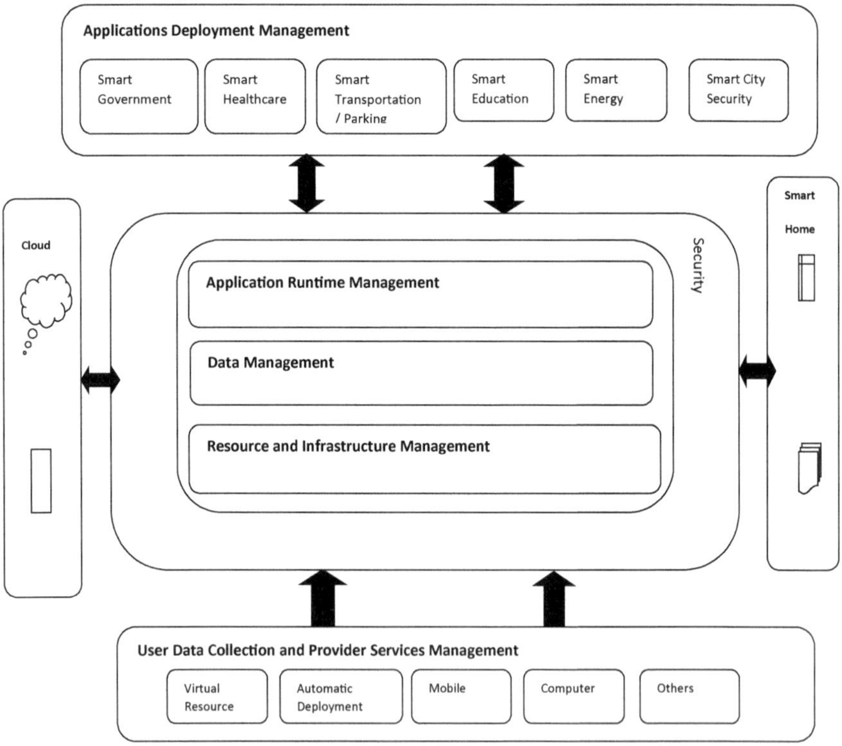

Fig. 1 The proposed architecture for making a smart city

will be collected using the existing communication services of GPS devices, cellular services (2G/3G/4G) of smartphones, and IoT [30, 31]. The collected data will then be processed and analyzed with the help of the existing semantic web technologies. Here, the main focus will be on deploying the architecture of smart cities on clouds which can further be used as SaaS [30, 31].

The proposed architecture of Fig. 1 will help the citizens of smart cities in their daily activities by sending time-to-time alerts and warnings to recall and remember day-to-day life-related things. The proposed system will act as an intelligent platform equipped with artificial intelligence techniques for people living in smart society. By combining data from different domains, this architecture will help in assisting citizens of smart cities in an intelligent manner, such as by sending alerts and warnings for their household items like for buying food items via a smart fridge.

The proposed architecture will help the drivers to take another route in case of traffic jam situations, automatically alert the heart patients if their heart bit rate crosses significantly over a threshold value while performing day-to-day activities. In this proposed smart city architecture, the raw data will be collected and processed to make it Internet-friendly, and then only it will be forwarded for uncertainty and usefulness checking. The new rules designed and implemented at this stage will be useful in describing the knowledge of the proposed model. The similar technique can also be used in describing the customized services, which will further provide 24 × 7 feedback to the citizens in the form of different types of alerts and specific warnings.

Figure 1 shows sensors, which will sense the raw data and this raw data will be transferred using communication services for performing further information processing tasks. Some of the important structures in which data collection becomes very important are tweets of different handles, text messengers, and database schemas, etc. The obtained formats will then be processed using semantic web technologies for converting them into a common structure. Here, the main objective is to convert the obtained different types of information into a commonly acceptable format called Resource Framework Description (RFD). This RFD will be used for interchanging information through webs, and it will facilitate heterogeneous data distribution and its integration for different types of smart cities. Further, different types of software applications will use RFF data for performing efficient and intelligent operations [32, 33].

The role of the common medium is very crucial in achieving the smart city goals. The currently available communication services, which are being frequently utilized in a smart city infrastructure are long-term evolution (LTE), 4G, Wireless fidelity (Wi-Fi), ZigBee, cable television, satellite communication, and worldwide interoperability for microwave access (WiMAX).

5 Discussions

There are various types of architecture that exists and describe smart city systems.

Figure 2 illustrates the energy management of the smart city as demonstrated by the Toshiba Group of Japan [34]. Some other groups like the Hitachi group in Japan have been actively working in the area of human care where environment-friendly smart cities will have low carbon emission.

Figure 3 demonstrates the following model for its remote communication services in Education & Healthcare [35]. This group has also shown a different look of the smart city where needed and required services in collaboration with private prop-

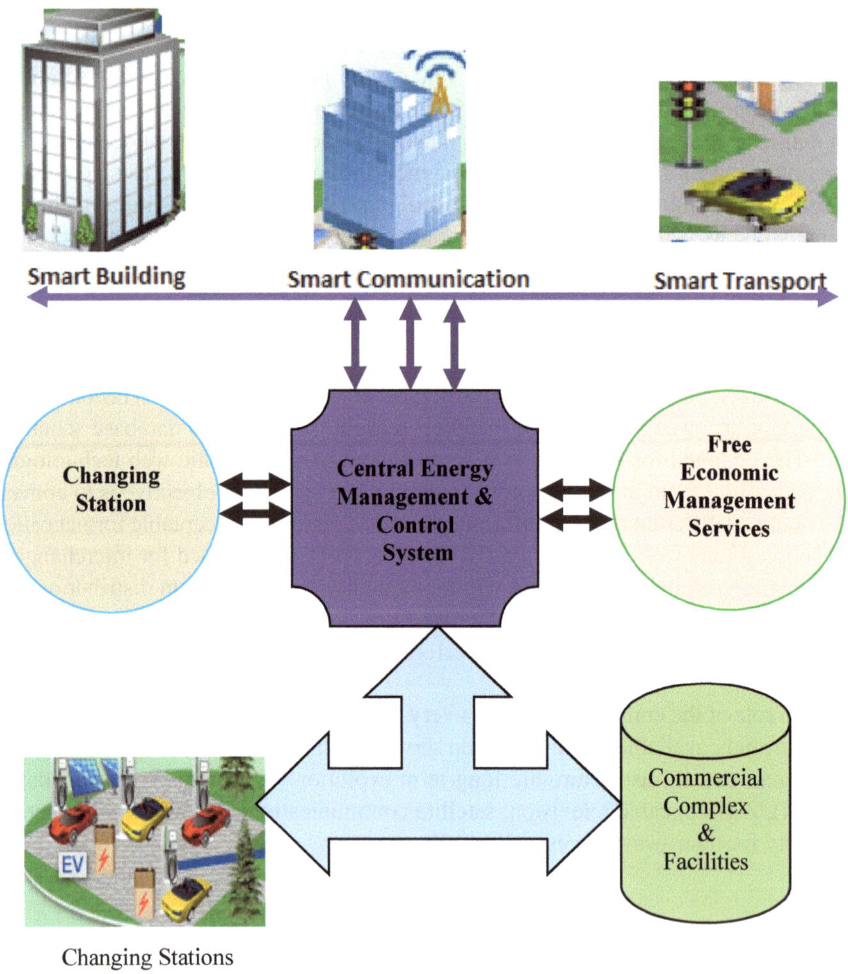

Fig. 2 Community energy management system of smart cities, drawn from data provided in [34]

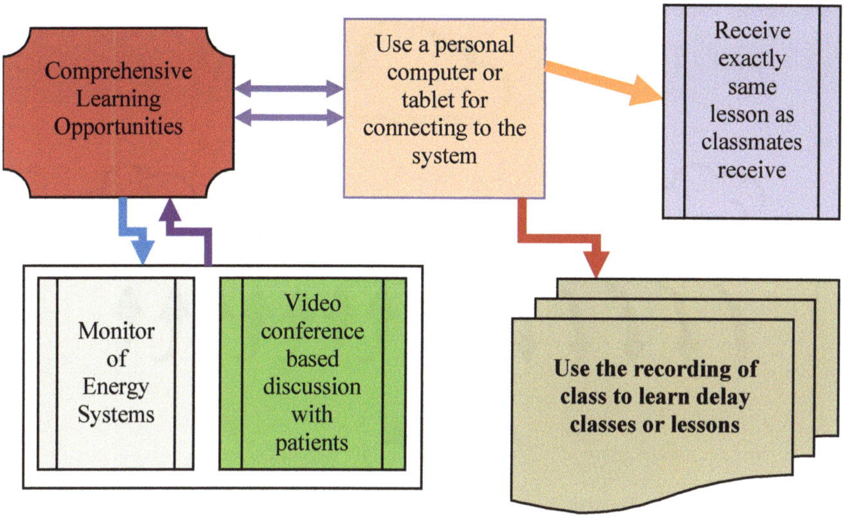

Fig. 3 Remote communication services for education and healthcare, drawn from data provided in [35]

erty are taken care of. Figure 3 demonstrates the methodology used by the Hitachi group, Japan that follows the shared use of neighborhood facilities. Becoming a smart city includes many aspects together with important characteristics like smart living, smart medical facility (e-Medical), smart home, smart buildings, smart transportation, smart water management system, smart waste management, smart energy (Renewable generation and storage) management, smart governance (e-governance), smart communication medium, smart AI-based networks, environmental awareness, and smart education (e-Education) [36, 37].

In order to make easy access to public services in smart cities, it has become necessary to visit the city centers or local centers in the suburb area. Because of the changing living conditions and expansion of cities the development has unfortunately led to increased distances to the service points. Due to IoT and ICT, it is now becoming easily possible to use the home delivery services for different items and things related to our daily life or even we can get certain services while traveling.

The information and communication technologies encourage and help the citizens to participate in the decision-making process of the country much more than before and it has become much more difficult for the authorities of different offices and organizations to keep their work behind closed doors. The people are able to directly interact with the government officials and elected representatives of our societies without any hesitation because of the easy accessibility of IoT and ICT. This interaction of people with officials in a smart city is presented in Fig. 4. Despite Sevier challenge of the digital divide, the availability and integration of ICT with the general public have brought a huge increase in the power of citizens. Hence, the meaning of remote geographical location is changing. Now, we can act globally and locally

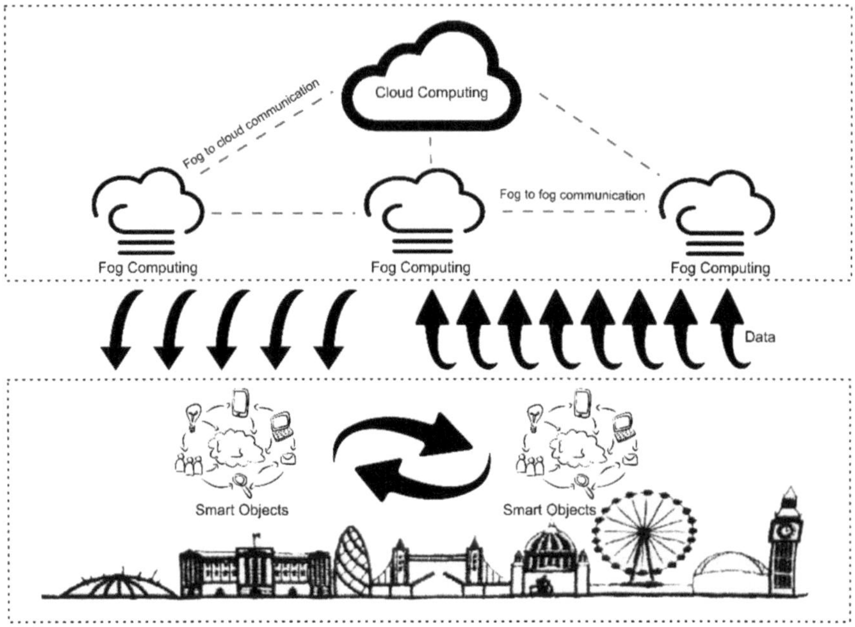

Fig. 4 Cloud computing, fog computing, IoT, and smart cities. Reprinted from Ref. [45], with kind permission from Springer Science+Business Media

while being in remote places. Further, at the same time, we can use the resources of places to which we do not even know [38, 39].

In the traditional distributed data management systems, the data gathering tasks were executed at a central location before the start of the data analysis. The results drawn were transmitted back to users as instructions after completing the data analysis. In the current era distributed systems, the data can be gathered in one place and can be analyzed at another place or same place. This permits the creation of a larger amount of input data and a much wider range of resources are required for data assessment and conclusions. Hence, the central decision-making system is not always required in the current era of a distributed environment. The open source/data systems permit easy availability of data and resources for analysis and further usage. The need for being smart in forthcoming societies will not be dependent on place/locality but it will depend on being connected to networks of societies. The properly integrated technological systems can be easily managed. Therefore, the whole world is considering it as an excellent opportunity for improving the quality of life through the development of smart cities on this earth. Hence, the smart city mission is now having a central position in urban development projects and is the center of attraction for all the countries [8].

The smart city projects require trillions of dollars of investments and these projects will provide excellent business opportunities for technology providers, investors, and the general public. Therefore, a new era of business with the development of smart cities in India will start with the commencement of smart cities. The new developments are merged with sensing, IoT and data monitoring technologies have now become key requirements for the efficient and real-time collection of metadata information from different sources which are enriched into city monitoring and operating systems through key performance indicators [40].

The new approaches to smart city concept include innovative views like lighting intelligent street lights with dimming control. The social technology will help in independent living. The detection algorithms will help in tracking the daily routines of citizens and offices and an alert can be generated for any suspicious behavior patterns. The parking sensors will detect the availability of spaces in a real-time environment to park the vehicles and traffic sensors will sense and provide space to drive ahead for motorists [25, 41]. Sourav et al. [42] highlighted the smart traffic management tool under the IoT framework. They used context aware traffic management algorithm for the removal of congestion.

The waste containers will have wireless sensors and therefore, the forecast of the fill level of these remote side containers will be easily managed from the offices. The citizens will inform the local authorities of repairing tasks and electrical faults and fires, etc. Further, the citizens will provide data to the concerned authorities for improving the efficient running of the city, e.g., cycle routes traveled, home and business energy meter readings and other serious issues. The ultimate vision of a smart city is to manage multiple systems appropriately at the city level with increased transparency, openness, and shared accountability. Therefore, it will help in creating a novel system which can further improve the outcomes and culture of a city [9].

The Government of India has taken initiatives that out of 100 proposed smart cities across states and union territories of India, only 20 would be selected this financial year 2017–18. The rest would join the club in 2 batches of 40 each in the next 2 years [15]. Table 1 describes the smart city projects of India and the corresponding key features.

The independent and error proof communication medium may play an important role in achieving the goals of the smart city concept in an actual sense. The existing communication services which are currently being utilized in a smart city are not sufficient [43, 44].

Hence, we need to further upgrade the currently being used for services with the help of artificial intelligence and soft computing techniques. The primary objective which can further improve the quality of life in the smart city is to connect all things related communication and information technologies (sensors and IoTs) that may help in increasing the comfort and safety levels of the life of citizens. An important example of this category is to provide a communication facility in the home domain for integrating the telephone and other communication systems including personal computers through the internet of things in a smart city. The need for the integration of a smart city with IoT and cloud computing is also conceptualized and discussed in [45], please see Fig. 4.

In the government sectors of many countries, the clouds and communication services are combined together with the help of AI approaches to obtain further

Table 1 Smart cities and their key features in India

S. No.	Name of city	Key features
1.	Lavasa (First fully planned hill city of India)	It is India's first planned city in hills since the independence It is a well-situated three hours drive from Mumbai, an hour drive from city Pune The Lavasa has 2/3 BHK flats and it is providing houses for socioeconomic classes It is supposed to lead the globe in hospitality, health, education, environment-friendly, and wellness The Lavasa has a permanent population of 0.3 million residents and many tourists come to this place for a visit
2.	Kochi (Kerala)	Smart City Kochi (SCK), a joint venture between Smart city Dubai and the Kerala government This project includes sustainability and environmental study, traffic impact study, urban design landscape guidelines, and strict plot development guidelines. This project is spread over 246 acres of land and it is predictable to create 90,000 direct jobs in the Indian market Smart city Kochi will probably claim to be providing the most advanced and reliable ICT infrastructure
3.	Haldia (West Bengal)	The EBTC (European Business and Technology Centre) is planning to start a pilot project for developing a smart city in Haldia of West Bengal, India. This project will focus on ecologically friendly environment and very low carbon emission will be in the footprint of the proposed city. The EBTC will help business units in India and Europe for fair and clean technology transfer projects In this smart city project, the Copenhagen Cleantech Cluster and EBTC will work together for the execution of work and they will provide research and innovation which are related to green technology
4.	Chennai (Metro City)	Metropolitan Water Supply and Sewerage Board migrated to an ERP platform to integrate discrete modules and enable MIS and citizen service complaints, billing and collection, and procurement leading to efficiency and transparency of operations

(continued)

Table 1 (continued)

S. No.	Name of city	Key features
5.	Bengaluru (Metro City)	It scores very good marks on smart city characteristics and it has smart people, and smart economy. But, the city needs rapid amendments to accomplish the criterion of other important factors like parking management, traffic management, waste management, energy management, and water management
6.	Mumbai (Metro City)	Municipal Corporation of Greater Mumbai has put in place a comprehensive ICT-enabled strategy for delivering citizen services through the corporation's portal and linked to SAP, which allows for real-time data and operations

better governance and control system. In the case of mobile health systems, the communication and information technologies are being frequently used to connect health care statistics, and location of patients from a remote source. Hence, the integration of smart cities with communication technologies and AI we can provide a further better safe and easily accessible infrastructure for improving the living standards and quality of life in smart cities.

With the help of forthcoming wireless technologies and wireless sensor networks, we can secure the future of Smart City systems.

6 Conclusions

The smart city concept is passing through an eco-friendly revolution and therefore, the smart city concept is entering into a new era where everything will become smart with the help of artificial intelligence technologies, sensor networks, and ICT. Further, we can aim to address some of the customized services in a smart city environment just by using semantic modeling. We aim to focus on the most important areas of the smart city environment. The metropolitan environment is in the decisive role when societies are facing the consequences of climate change and thinking not to destroy the natural things. The urbanization is an upcoming and fast-growing trend in the world. The smart systems and their integration with other artificially intelligent systems need to be developed for providing better services to the people, handling the population growth pressure, and bringing down the impacts of global warming on smart cities.

The significant pressure is regularly increasing on all of us to decrease the environmental impact. The sustainable transformation of cities is the only possible choice if it is done in a smart way where nature and environment-based things are untouched while developing smart cities. The smart city design, operation, and management needs to be done at the system level where environment growth should be given the highest priority. The traditional sector-based industries and value chains are also changing with the change of time and business opportunities. Further, completely new business models have been started by newcomers which are helping the environment to grow in its natural way, whereas the radical inventions and paradigm shifts are having a high impact on the environment. Therefore, the lives of metro cities are changing at an unimaginable rate where it has become almost impossible to get fresh air for breathing in metro cities.

The future work of this chapter is to perform some onsite experiments on the proposed ideas which have been discussed. It may include the discovery of different types of real-time heterogeneous information and proposing a realistic semantic knowledge model for combining sensor data of smart cities for analysis purpose. The data interoperability and scalability aspects can also be included in the forthcoming architecture of the smart city.

References

1. Allwinkle, S., Cruickshank, P.: Creating smarter cities: an overview. J. Urban Technol. **18**(2), 1305–1453 (2011)
2. Baqir, M.N., Kathawala, Y.: Ba for knowledge cities: a futuristic technology model. J. Knowl. Manag. **8**(5), 83–95 (2004)
3. Caragliu, A., del BoC, Nijkamp P.: Smart cities in Europe. J. Urban Technol. **18**(2), 65–82 (2011)
4. Chourabi, H., Nam, T., Walker, S., Gil-Garcia, J.R., Mellouli, S., Nahon, K., Scholl, H.J.: Understanding smart cities: an integrative framework in system science. In: 45th Hawaii International Conference, HICSS 2012, pp. 2289–2297. IEEE (2012)
5. Cocchia, A.: Smart and digital city: a systematic literature review. In: Dameri, R.P., Sabroux, C. (eds.) Smart City: How to Create Public and Economic Value with High Technology in Urban Space. pp. 13–43. Springer International Publishing, Switzerland (2014)
6. Marinova, D., Philimore, J.: Models of innovation. In: Shavinina, L.V. (ed) The International Handbook on Innovation, pp. 44–53. Elsevier (2003)
7. Schuler, D.: Digital cities and digital citizens. In: Tanabe, M., van den Besselaar, P., Ishida, T. (eds.) Digital Cities II: Computational and Sociological Approaches, pp. 71–85. Springer, Heidelberg (2002)
8. Harrison, C., Eckman, B., Hamilton, R., Hartswick, P., Kalagnanam, J., Paraszczak, J., Williams, P.: Foundations for smarter cities. IBM J. Res. Dev. **54**(4), 350–365 (2010)
9. Yoshihito, Y., Sato, Y., Hirasawa, A., Takahashi, S., Yamamoto, M.: Hitachi's vision of the smart city. Hitachi Rev. **61**(3), 111–118 (2012)
10. Faisal, R.: Spamming the internet of things: a possibility and it's probable solution. In: The 9th International Conference on Mobile Web Information Systems. Procedia Computer Science, vol. 10, pp. 658–665 (2012)
11. Anthopoulos, L., Fitsilis, P.: From online to ubiquitous cities: the technical transformation of virtual communities. In: Sideridis, A.B., Patrikakis, C.Z. (eds.) Next Generation Society: Technological and Legal Issues. Proceedings of the Third International Conference, eDemocracy, Athens, Greece, vol. 26, pp. 360–372 (2009)
12. Borja, J.: Counterpoint: intelligent cities and innovative cities. Universitat Oberta de Catalunya (UOC) Papers. E-J. Knowl. Soc. **5**, 1–12 (2007)

13. Edvinsson, L.: Aspects of the city as a knowledge tool. J. Knowl. Manag. **10**(5), 6–13 (2006)
14. Klein, C., Kaefer, G.: From smart homes to smart cities: opportunities and challenges from an industrial perspective. In: Proceedings of the 8th International Conference, NEW2AN and 1st Russian Conference on Smart Spaces, ruSMART 2008, pp. 260–270. St. Petersburg, Russia (2008)
15. Jennings, P.: Managing the risks of Smarter Planet solutions. IBM J. Res. Dev. **54**(4), 1–9 (2010)
16. Moser, M.A.: What is smart about the smart communities movement?. Electron. J. **10**(1), 1–11 (2001). http://www.ucalgary.ca/ejournal/archive/v10-11/v10-11n1Moser-print.html. Last accessed 15 July 2017
17. Urbantide. (n. d.) Overview of smart cities maturity models. https://static1.squarespace.com/static/5527ba84e4b09a3d0e89e14d/t/55aebffce4b0f8960472ef49/1437515772651/UT_Smart_Model_FINAL.pdf. Last accessed 1 Apr 2017
18. Gaur, A., Scotney, B., Parr, G., McClean, S.: Smart city architecture and its applications based on IoT. Procedia Comput. Sci. **52**, 1089–1094 (2015)
19. Lee, J.H., Hancock, M.G., Hu, M.: Towards an effective framework for building smart cities: Lessons from Seoul and San Francisco. Technol. Forecast. Soc. Chang. **89**, 80–99 (2014)
20. Mell, P., Grance, T.: The NIST Definition of Cloud Computing. National Institute of Standards and Technology, Version 15 (2009)
21. EPIC: Smart city information architecture and functional platform (2011). http://www.epic-cities.eu/sites/default/files/documents/D3.2%20Smart%20City%20Info%20Architecture.pdf. Last accessed 15 Apr 2017
22. Shukla, P.: Smart cities in India (2015). http://terienvis.nic.in/WriteReadData/links/Smart%20Cities%20in%20India_Report_pagewise-5937837909069130880.pdf. ENVIS Centre on Renewable Energy. Last accessed 21 May 2017
23. European smart cities: Smart cities final report (2017). http://www.smart-cities.eu/download/smart_cities_final_report.pdf. Last accessed 01 June 2017
24. VTT: Smart city research highlights (2015). http://www.vtt.fi/inf/pdf/researchhighlights/2015/R12.pdf. Last accessed 01 June 2017
25. Yousif, A., Farouk, M., Bashir, M.B.: A cloud-based framework for platform as a service. In: International Conference on Cloud Computing (ICCC), Riyadh, pp. 1–5 (2015)
26. Alam, A.F.B., Soltanian, A., Yangui, S., Salahuddin, M.A., Glitho, R., Elbiaze, H.: A cloud Platform-as-a-Service for multimedia conferencing service provisioning. In: IEEE Symposium on Computers and Communication (ISCC), Messina, Italy, pp. 289–294 (2016)
27. Doukas, C., Antonelli, F.: A full end-to-end platform as a service for smart city applications. In: IEEE 10th International Conference on Wireless and Mobile Computing, Networking, and Communications (WiMob), Larnaca, pp. 181–186 (2014)
28. Ribas, M., Sampaio, L.A., Neuman de Souza, J., Rubens de Carvalho Sousa, F., Oliveira, M.L.: A platform as a service billing model for cloud computing management approaches. IEEE Lat. Am. Trans. **14**(1), 267–280 (2016)
29. Hong, L.T., Schahram, D., Georgiana, C., Alessio, G., Waldemar, H., Duc, H.L., Daniel, M.: CoMoT—a platform-as-a-service for elasticity in the cloud. In: IEEE International Conference on Cloud Engineering (IC2E), Boston, MA, pp. 619–622 (2014)
30. Sami, Y., Pradeep, R., Ons, B., Roch, H.G., Monique, J.M., Paul, A.P.: A platform as-a-service for hybrid cloud/fog environments. In: IEEE International Symposium on Local and Metropolitan Area Networks (LANMAN), Rome, Italy, pp. 1–7 (2016)
31. Graubner, P., Baumgärtner, L., Heckmann, P., Müller, M., Freisleben, B.: Dynalize: Dynamic analysis of mobile apps in a platform-as-a-service cloud. In: IEEE 8th International Conference on Cloud Computing, New York City, NY, pp. 925–932 (2015)
32. Dhuldhule, P.A., Lakshmi, J., Nandy, S.K.: High-performance computing cloud—a platform-as-a-service perspective. In: International Conference on Cloud Computing and Big Data (CCBD), Shanghai, pp. 21–28 (2015)
33. Vanhove, T., Vandensteen, J., Seghbroeck, J.V., Wauters, T., De Turck, F.: Kameleo: design of a new platform-as-a-service for flexible data management. In: IEEE Network Operations and Management Symposium (NOMS), Krakow, pp. 1–4 (2014)

34. Toshiba Group: Annual report—environmental report (2013). https://www.toshiba.co.jp/env/en/communication/report/index.htm
35. Yoshikawa, Y., Hanafusa, Y., Matsuda, T., Hirayama, I.: Service infrastructure for next-generation smart cities. Hitachi Rev. **61**(3), 119–125 (2012)
36. Seelam, S.R., Dettori, P., Westerink, P., Yang, B.B.: Polyglot application auto scaling service for platform as a service cloud. IEEE International Conference on Cloud Engineering (IC2E), Tempe, AZ, pp. 84–91 (2015)
37. Baek, S., Kim, K., Altmann, J.: Role of platform providers in service networks: the case of Salesforce.com app exchange. In: IEEE 16th Conference on Business Informatics, Geneva, pp. 39–45 (2014)
38. Bobák, M., Hluchý, L., Tran, V.: Tailored platforms as cloud service. In: IEEE 13th International Symposium on Intelligent Systems and Informatics (SISY), Subotica, pp. 43–48 (2015)
39. Giessmann, A., Kyas, P., Tyrväinen, P., Stanoevska, K.: Towards a better understanding of the dynamics of platform as a service business models. In: 47th Hawaii International Conference on System Sciences, Waikoloa, HI, pp. 965–974 (2014)
40. Kolb, S., Wirtz, G.: Towards application portability in platform as a service. In: IEEE 8th International Symposium on Service Oriented System Engineering (SOSE), Oxford, pp. 218–229 (2014)
41. VTT: Productivity Leap with IoT (2013). https://www.vtt.fi/inf/pdf/visions/2013/V3.pdf. Last accessed 3 May 2017
42. Sourav, B., Chinmay, C., Sumit, C.: A survey on IoT based traffic control and prediction mechanism. In: Internet of Things and Big data Analytics for Smart Generation (Intelligent Systems Reference Library), Chap. 4, vol. 154, pp. 53–75. Springer (2018)
43. Kamta, N.M.: Secure cloud-based validation and admittance control mechanism for supervising transmission services. Int. J. Intell. Transp. Syst. Res., 1–18 (2018)
44. Simon, P.A.: Artificial Intelligence and Internet of Things in the Development of Smart Sustainable Cities, Adoption of Circular Economies in the 4th Industrial Revolution and 8th Green Standards Week, Zanzibar International Telecoms Union (ITU), pp. 1–38 (2017)
45. Hosseinian-Far, A., Ramachandran, M., Slack, C.L.: Emerging trends in cloud computing, big data, fog computing, IoT and smart living. In: Dastbaz, M., Arabnia, H., Akhgar, B. (eds.) Technology for Smart Futures. Springer, Cham (2018)

Dr. Kamta Nath Mishra was born on August 15, 1973, in Kushinagar district of Uttar Pradesh, India. He received his Bachelor of Science (B.Sc., Maths) degree from the University of Gorakhpur, India, in 1992 and Master of Computer Application (MCA) degree from Madan Mohan Malviya Engineering College (currently MMMUT), Gorakhpur, UP, India in 1996. Dr. Mishra completed his M.Tech. (Software Systems) degree from Birla Institute of Technology and Science (BITS) Pilani, India in 2003 and Ph.D. (Engg.) from the CSE Department of B.I.T. Mesra–India, in May 2015. Dr. Mishra has more than 17 years of teaching and research experience. Currently, he is working as an Assistant Professor (Senior Grade) at Department of CS&E, B.I.T Mesra, Ranchi, India since August 2009. He has worked as a faculty member in the Department of Computer Science, Joint program of Michigan State University the USA and University of Sebha, Libya, from October 2006 to July 2009. He was a senior lecturer at B.I.T Mesra, (Noida Campus) from July 2004 to September 2006. Dr. Mishra has worked as a senior project engineer from September 2003 to June 2004 and project engineer from September 2000 to August 2003, in the Centre for Development of Advanced Computing (Ministry of Communication & IT, Government of India) Noida, Uttar Pradesh. Before joining CDAC, Dr. Mishra worked as a lecturer in CS&E Department at Krishna Institute of Engineering & Technology (KIET), Ghaziabad, India, from July 1998 to August 2000. Dr. Mishra has published more than 30 research papers in journals and conferences of international repute. His research interest includes Biometric Systems, Image Processing, Analysis of Algorithms and Distributed Cloud Computing. Dr. Mishra is a professional member of IEEE Biometric Society USA, and ACM, USA.

Web page: https://scholar.google.co.in/citations?user=-K2qSkoAAAAJ&hl=en

LinkedIn: https://www.linkedin.com/in/dr-kamta-nath-mishra-2a274a20/

Dr. Chinmay Chakraborty is an Assistant Professor (Senior Grade) in the Department of Electronics and Communication Engineering, Birla Institute of Technology, Mesra, India. His primary areas of research include Wireless Body Area Network, Internet of Medical Things, Energy-Efficient Wireless Communications and Networking, and Point-of-Care Diagnosis. Prior to BIT, he worked at the Faculty of Science and Technology, ICFAI University, Agartala, Tripura, India as a senior lecturer. He worked as a Research Consultant in the Coal India project at Industrial Engineering & Management, IIT Kharagpur. He worked as a project coordinator of Telecom Convergence Switch project under the Indo-US joint initiative. He also worked as a Network Engineer in System Administration at MISPL, India. He received his Ph.D. in Electronics & Communication from BIT Mesra, India. He is the author of PSTN-IP Telephony Gateway for Ensuring QoS in Heterogeneous Networks (2014). He is an Editorial Board Member in the Journal of Wireless Communication Technology and also a member of the International Advisory Board for Malaysia Technical Scientist Congress and the Machine Intelligence Research Labs. He is an Editorial Board Member in the Journal of Wireless Communication Technology and also a member of the International Advisory Board for Malaysia Technical Scientist Congress and the Machine Intelligence Research Labs. He has been also editing three books on Springer (Advances of IoT in Computational Intelligence and Bioinformatics), IGI (Advanced Classification Techniques for Healthcare Analysis), and IET (Telemedicine for Health Monitoring: Advances, Technologies, Design, and Applications) and Guest Editor of the Special Session "Next Generation Networks for Smart Healthcare" ICICC-2019. He got Outstanding Researcher Award from TESFA, 2016 and also got Young Faculty Award from VIFA 2018.

Web page: https://sites.google.com/view/dr-chinmay-chakraborty

LinkedIn: https://www.linkedin.com/in/dr-chinmay-c-35017310/

The Future of Mobility with Connected and Autonomous Vehicles in Smart Cities

Patrice Seuwou, Ebad Banissi and George Ubakanma

Abstract Cities around the world are being wrecked by the ever-increasing burden of traffic. Smart cities are a recent innovation perceived as a winning strategy to cope with some severe urban problems such as traffic, pollution, energy consumption, waste treatment. This concept is attracting significant interest in the world of technology and sensors. Governments can streamline the way cities are run, saving money and making them more efficient as a result. Rapid urban developments, sustainable transportation solutions are required to meet the increasing demands for mobility whilst mitigating the potentially negative social, economic and environmental impacts. This study analyses the smart mobility initiatives and the challenges for smart cities with connected and autonomous vehicles (CAVs), and it also highlights the literature that supports why CAVs are essential for smart maintainable development as part of the intelligent transportation system (ITS).

Keywords Smart cities · Smart mobility · ITS · IoT · Security · Connected and Autonomous Vehicles (CAVs)

1 Introduction

For the majority part of the twentieth century, the concept of a smart city was only science fiction pictured by the popular media. But very recently with considerable progress achieved in the development of computing and electronic devices, the vision that an entire city could be transformed into a smart town is becoming a reality [1]. The idea of the smart city captured most people attention during the last decade as a blend

P. Seuwou (✉) · E. Banissi · G. Ubakanma
Division of Computer Science and Informatics, School of Engineering, London South Bank University, London, UK
e-mail: seuwoup@lsbu.ac.uk

E. Banissi
e-mail: banisse@lsbu.ac.uk

G. Ubakanma
e-mail: george.ubakanma@lsbu.ac.uk

M. Farsi et al. (eds.), *Digital Twin Technologies and Smart Cities*, Internet of Things, https://doi.org/10.1007/978-3-030-18732-3_3

of beliefs on how technology, in general, could be used to transform how cities around the world work, while improving their competitiveness, offering new ways of solving problems linked to poverty, social deprivation, pollution and poor environmental issues [2]. Smart cities are often seen as collections of intelligent devices installed across the city able to communicate with each other while providing constant data on the movement of people and objects. Over 50% of the world population now live in cities and it is expected that by 2050, cities will be home for about two-thirds of the world inhabitant [3]. As the population in cities continue to rise, the need for mobility as well as its burdens on the environment, social stability and the economy will grow rapidly [4]. People are attracted to cities mainly because of all the great opportunities they offer. In cities, people are able to live and work, companies are able to settle, grow and recruit competitive staffs, young people able to go to schools and universities. However, cities are also places for diseases mainly because in cities there are high volumes of cars, traffic, CO_2 emission, high cost of living where waste production and pollution are worse. The possible applications of smart vehicles such as connected vehicles (Car2X technologies), autonomous vehicles (AVs) as well as connected and autonomous vehicles (CAVs) are wide-ranging, spanning on a variety of different sectors. CAVs appear to be a possible answer to contemporary transportation problems. Mass adoption of this emerging technology as a mode of transportation will reduce issues linked to emissions and energy consumption, while improving traffic flow, accessibility and efficiency of transportation systems, road safety and city efficiency among other benefits [5–8]. The deployment of CAVs will provide a time and space for other activities to take place from catching up on emails to watching TV. This chapter defines the concept of smart cities, it analyses connected autonomous vehicles (CAVs) as a prospective future mobility solution for smart and sustainable development. We will also identify challenges and security threats of CAVs as critical risks to the expansion of smart and sustainable cities around the world.

Typical CAVs are equipped with technologies such as Lidar, video camera, positioning estimator, distance sensors. These vehicles can "talk" to each other exchanging information such as vehicle size, position, speed, heading and turn signal status. Generally, video cameras are mounted near the rear-view mirror, the camera detects traffic lights and any moving objects, the Lidar positioned on the roof of the car as a rotating sensor scans the area in a radius of 60 m for the creation of a dynamic, three-dimensional map of the environment. A position estimator, which is a sensor mounted on the left rear wheel measures lateral movements and determines the car's position on the map. Distance sensors are made up of four radars, three in the front bumper and one in the rear bumper, measure distances to various obstacles and allow the system to reduce the speed of the car. CAVs will also be equipped with an event data recorder or EDR also referred to as an automotive black box, recording information related to vehicle crashes or accidents.

The main contributions of this paper are as follows: (1) We present CAVs as the way forward and future for the intelligent transportation system. (2) We analyse the capabilities of this technology to optimise network capacity, reduce congestion, make people's journeys stress free and increase safety and reduce pollution. (3) We explore the main challenges CAVs are facing while highlighting the key literature to encourage dialogue and engagement with national and local government, network operators, the automotive industry, technology providers, the logistics sector, etc.—as well as all the different stakeholders for who the road network is fundamental to connecting people and places.

The rest of the paper is organised as follows. In Sect. 2, we explore the concept of smart city and its challenges. In Sect. 3, we discuss how sustainable mobility could be reinvented for smart cities. Section 4 evaluate how transportation could become eco-friendly. Section 5 review the impact of CAVs on KPIs for smart cities. Section 6 explores the barriers to CAVs implementation and Sect. 7 presents our conclusion and future work.

2 Smart City Concept and Challenges

In the early 1990s, the expression "smart city" was coined to indicate a city that has been transformed to a modern urban landscape with the effect of globalisation, extensive usage of technology and innovation [9]. In the past few years, this concept has attracted significant attention in the context of urban development policies and from various governments interested in collecting more and more data about their population. In this setting, security agencies, law enforcement organisations, secret services and other relevant bodies will be able to monitor, collect and analyse data about the movement of people going to school, work, libraries, hospital and other community services, goods, traffic information, power plants activities, waste management, water supply networks, energy facilities in real time [10]. It is very important to recognise that infrastructure is a vital element for smart cities. Technology is one of the tools that make it possible but fundamental for the city to be truly smart, there should be a connection, combination and integration of all parts of the puzzle. For cities to gradually assume a critical role of leaders in innovation in sectors such as business, transportation, health in the digital economy, e-services enabled by Internet and broadband, network technologies are very important. Around the world, as cities continue to grow, more and more people are pursuing better lifestyle, challenges related to economic development, population growth and social progress seriously need to be considered carefully. In the reviewed literature, challenges have been identified and classified into six main city dimensions: Governance, Economy, Mobility, Environment, People and Living [11]. They represent the specific aspects of a city upon which smart initiatives impact to achieve the expected goals of a smart city strategy (sustainability, efficiency and high quality of life). Addressing the problems and development priorities of cities in a global and innovation-led world is the most important challenge of smart cities. The section below shows a possible classification of challenges in European cities:

- **Governance** (Flexible governance, shrinking cities, territorial cohesion, combination of formal and informal government).

- **Economy** (Unemployment, shrinking cities, economic decline, territorial cohesion, mono-sectorial economy, sustainable local economies, social diversity as source of innovation, ICT infrastructure deficit).
- **Mobility** (Sustainable mobility, inclusive mobility, multimodal transport system, Urban ecosystems under pressure, traffic congestion, non-car mobility, ICT infrastructure deficit).
- **Environment** (Energy saving, shrinking cities, holistic approach to environmental and energy issues, urban ecosystems under pressure, climate change effects, urban sprawl).
- **People** (Unemployment, social cohesion, poverty, ageing population, diversity as source of innovation).
- **Living** (Affordable housing, social cohesion, health problems, emergency management).

Governments around the world who aspire to develop smart cities really need to change their governance models. Reduction of greenhouse gases emission, sustainable development, improvement of the energy efficiency of urban infrastructures are some other societal issues that should be addressed. In smart mobility, the overall challenge is to accomplish an inclusive, sustainable and efficient transportation system of people and products. This could be achieved with the introduction of CAVs. On one hand, deploying a multimodal public transport system also known as a combined transport system, making public transport accessible to all people and encouraging alternatives to car-based mobility are the three focus points that will allow improvement of connectivity, reduction of congestion and pollution in cities [11]. On the other hand, as a follow-up to the Paris Agreement adopted by consensus in 2015 on the necessity for countries to deal with greenhousegas emission and achieve a sustainable development, there is a rising ecological demand for cities around the world to find ways of reducing energy consumption, CO_2 emissions and pollution. Cities able to prioritise those crucial aspects and overcome the challenges listed above are crucial elements that set apart cities when it comes to good quality of life.

3 Smart Sustainable Mobility for Smart Cities

During the last 50 years, the sizes of cities have been growing considerably all over the world and it is predicted that city dimensions will increase even more in the future. In the past, to resolve the growing urban environment needs, most governments used to build more road and streets. Today, due to the lack of public funds and physical space, the transport network with a reduced capacity and the increased size of the population, is becoming overloaded, creating more congestion, CO_2 emissions and serious disruptions to citizens [12].

Mobility, as we know it today, must be carefully evaluated, considered and reinvented. At this moment in our development, we are at a junction for a paradigm shift that will encourage the increased use of smart mobility services with ultra-efficient vehicles, renewable energy, cooperative systems, innovation and optimization of cities resource allocation with ITS. In a bid to deliver better services to the citizens and increase the quality of life in urban spaces, smart cities are considered to be a winning strategy to enhance environmental quality [13]. Mobility is recognised as one of the most important elements to support the functioning of the urban area for a better quality of life. Yet, major issues such as traffic jam, the time it takes to cross the city, the bad state of work–life balance, street congestion and the excessive price of public local transport services are only some of the negative impacts and problems it produces. Smart mobility with CAVs appears to be one of the most promising aspects of smart cities with the capability to substantially improve the quality of life of most city workers and the stakeholders [14]. The following six categories summarise the most important smart mobility objectives retrieved from existing literature [15, 16].

1. Reducing pollution;
2. Reducing traffic congestion;
3. Increasing people safety;
4. Reducing noise pollution;
5. Improving transfer speed;
6. Reducing transfer costs.

Additionally, smart mobility systems include a range of interventions with AVs and CAVs that must be viewed through a wide-angle lens as a technology involving several disciplines (psychologists, sociologists, computer scientist and engineers). These systems use all the paradigms comprising the smart city, which includes digital city, green city and knowledge city. It also includes both actions concerning the mode of transport which affect the behaviours of residents and the introduction of vehicles with certain characteristics. Reference [17] portrayed the AV emphasising on key concepts such as "Connected" and "Big Data". Indeed, in the future, we will not only have connected vehicles better known in the research community as VANET (Vehicular Ad hoc Network) able to talk with each other (V2V) and to infrastructures (V2I). These vehicles will be autonomous equipped with internet and communication capabilities as part of the internet of things, ensure communication between all contributing agents and stakeholders such as pedestrians, road infrastructure (road side base stations or intelligent traffic lights…), authorities, other vehicles.

With ITS and the increased integration of information and communication technology, it is beyond doubt the question now is not whether the whole transportation system will be transformed and revolutionised, but how soon would this happen. Smart technologies will be introduced at all stages of the systems including vehicles, traffic light, surrounding infrastructure, systems management, energy supply and the delivery of services around the city. This trend continuing gradually will form a powerful intelligent ecosystem, on one hand, part of the internet of things (IoT) where objects and devices inevitably collect and share that data that they receive, and in another hand, part of the Internet of Everything (IoE) where cities are able to bring together data collected from people, processes and things in a bid to make networked connections more relevant than ever before, turning eventual random data into information and knowledge to create richer experiences and unprecedented economic opportunities for individuals, businesses or even countries. Autonomous vehicles equipped with communication facilities will have the ability to monitor and collect valuable data about their surroundings and provide important information to other road users and infrastructure systems. The IoT and IoE will make communication between all pieces of the puzzle possible by enabling a whole range of service automation and optimization. As part of the solutions proposed in existing smart mobility literature, it appears that concepts such as car sharing may become very popular. It is expected that CAVs will be electric. Car sharing is a service that allows you to use a car reservation, picking it up and bringing it back to a parking lot and paying due to the use made. It allows reduction of urban congestion, reduction of polluting emissions (gas and noise), reduction in employment of public space and in general, a new push towards the use of public transport [18]. TESLA company with their new business model is one of the leading car manufacturers developing Electric Vehicles (EVs) equipped with powerful, high performing, reliable and cost-effective batteries. These initiatives are part of the movement to combat climate change. Eventually, new vehicles producers will innovate and promote green and sustainable mobility. As a strategy to gradually migrate from conventional vehicles running on gasoline or diesel, car manufacturers may develop hybrid vehicles at the beginning and progressively move to electric vehicles powered with batteries. This move will ultimately disrupt several sectors and have a huge impact on the oil industry. On one hand, authorities should put in place infrastructure and policies supporting sustainable mobility such as the creation of bicycle lanes or interventions aimed at changing mobility as the creation of restricted traffic zones. The expansion or creation of bicycle lanes is an intervention that is closely linked to the use of the bicycle as a mean of private transport and could have positive effects on the spread of bike sharing; on the other hand, a series of integrated policies that can be implemented to change the mobility system, in particular by the public decision maker (for example, incentives for the use of less polluting fuels, tax incentives or measures such as higher taxation on polluting fuels). One efficient way to achieve lower greenhouse gas emissions is by Intelligent Transport Systems and Services (ITS) that utilises information and communications technologies. ITS helps achieve transport policy goals by shifting the focus from expensive transport infrastructure construction towards efficiency and fluency of mobility and logistics while creating and enabling new business. It must be noted that the introduction of ITS technologies will also significantly contribute to improved efficiency, safety, environmental impact and overall productivity of the transportation system.

4 Eco-Conscious Provisions for Public Transportation

Saying that we are connected to the World Wide Web (WWW) generally means to people that we are using our smart phones, smart watches, tablets or computers to access the internet. According to [19], in the year 2008, the number of things connected to the internet somehow surpassed the number of people on earth. By

2020, it is estimated that the number of items and things connected to the net around the world will be over 50 billion shaping a rich digital environment. These elements will be shaping the very fabric of our digital culture with numerous sensors embedded to our mobile devices storing and sharing data about our lives. In the future, smart cars will also be part of this ecosystem. South Korea is known as the most wired country on earth with a fast Internet connection and an impressive broadband connection even superior to the one currently in the UK. People spend a considerable amount of time on the internet. In some places, the internet is now classified as an addiction. More and more people are spending a considerable part of their lives online. With intelligent transportation systems in smart cities, vehicles will be mobile platforms more like a computer on the road or rather computer network on the road together with huge scale cloud infrastructures and other network-enabled devices. Vehicles will have the ability to cooperate and interact with each other and roadside base stations, therefore, creating value across numerous sectors in smart cities.

While being driven by autonomous technologies, people will be able to read, watch movies, play games, sleep or work and be more productive. As part of the process to realise the truly sustainable smart city, vision technologies around the world should focus their energy in developing applications and devices with IoT capabilities. Enabling things and objects such as street cameras, traffic jam control systems, sensors for transportation times to be smarter. This will then give access to applications' developers access to these devices data through Application Programming Interface (API) technologies.

In most advanced European cities, transportation is regarded as one of the main activities for daily living. Most commuters spend on average an hour or more per day travelling [20]. London is a multicultural city and many people work in London but live outside the city mainly because of the high prices of accommodation. The city aims at providing various modes of transportation (e.g., buses, trains, boat, trams, metros, "rentable" bikes and flying car in the future) while considering the environmental effects given that 12% of global CO_2 emissions are caused by transportation means [21]. In a sustainable global system, the three pillars of sustainable development are Environmental, Economic and Social [22]. According to a study of UK air quality [23], road pollution is more than twice as deadly as traffic accidents, while car pollution causes severe heath damage and risks in premature deaths [24]. On the other hand, one needs to take into consideration the emerging landscape: cities going digital by deploying various sensors and additional information is provided by individuals through their mobile devices. In this context, IoT as an underlying technology aims at creating smart environments/spaces for energy and mobility (as described by the European Research Cluster on the Internet of Things [25].

5 Green Mobility and the Impact of CAVs on KPIs

There are multitude possible applications of CAVs across several sectors. This may include military to save soldiers in dangerous combat zones without risking more human lives, in the marine for example to search missing planes and other precious items in dangerous deep sea, aerial, this is already been used, for example armed drones (remotely piloted aircraft) have been used by the US and UK forces to carry out strikes in area such as Afghanistan to locate and eliminate terrorists [26]; space where unmanned spacecrafts could be launched in deep space to explore and seek new life in faraway locations where no man has gone before [27, 28] public roads, private and public transportation, to children and, elderly passengers, disabled people and the public in various location within the city. The technology can also be used in warehousing, in agriculture, with drones able to inspect and monitor crops and other resources, working in dangerous and unsafe environments such as nuclear facilities or locations with landmines. It can also be used to deliver humanitarian aid to populations in disaster zones around the world. The British government is actively considering the potential for the UK to adopt this technology particularly in the roads sector [29]. As CAVs become a reality, they will have a growing impact on a range of Key Performance Indicators (KPIs) measured by cities and road authorities as indicated in the section below [30]:

- **Journey time reliability**: Due to the CAVs ability to drive closer together, the impact caused by congestion will be reduced. The adoption of CAVs will increase roadway capacity without impacting on safety since machines can keep minimum distances and still drive safely when compared with human drivers. The journey time reliability will be expected to increase as incidents and accidents will be reduced.
- **Traffic volume**: It is still not very clear what is going to be the impact of CAVs on traffic volumes, but it is believed that CAVs will lead to a huge reduction of vehicles on the road network because of the car share scheme. At the same time there may also be an increase in the traffic on the roads due the opportunity CAVs provides to children, elderly and disables passengers.
- **Road safety**: It is believed that 90% of all accidents are caused by driver error, therefore, by handing driving duties over to computers and technologies, it is believed that the number of crashes will significantly decrease. And where collisions do occur, their severity rate is expected to be reduced as CAVs will be able to react quicker than the average human driver, thus mitigating the severity of the collision.
- **Safety of the most vulnerable road users**: CAVs will be developed with advanced technology devices such as Lidar, sensors, camera and several processors able to predict vulnerable road users' actions, therefore it will improve safety of the most vulnerable road users (children, disables, elderly people…)
- **Ensuring the road network supports economic growth potential**: By reducing congestion on the road network and improving journey time reliability, the road

network will support the economic growth potential of an area by allowing efficient and reliable mobility.
- **Reduce carbon emissions associated with road traffic**: On a per vehicle basis, carbon emissions will be expected to fall as CAVs are adopted as the technology which will improve driving efficiency (for example reducing stop/start driving conditions). However, if there is an overall increase in traffic on roads, then aggregated carbon emissions may remain static or increase.
- **Reduce the negative impact of road traffic on local air quality**: As with carbon emissions, on a per vehicle basis, local air quality conditions will be expected to improve. As Connected and Autonomous Vehicles are adopted, the technology will improve driving efficiency. However, if there is an overall increase in traffic on roads then local air quality conditions may remain static or worsen.
- **An accessible and integrated road network that provides equal opportunity for use**: CAVs will open up the road network for equal opportunity use. This will increase mobility options and travel horizons for large sections of the population, resulting in increased economic, social and well-being opportunities.
- **Freight optimisation**: From connected platooning to automated and predictable last mile deliveries, CAVs will have a role to play in optimising and streamlining logistic movements. This, in turn, will help to improve the ability to both schedule and meet reduced delivery times, helping improve customer loyalty and satisfaction.
- **Increase the number and proportion of people using active modes of travel (walking and cycling)**: The impact of CAVs on the number of people using active modes of travel is unknown. Persons currently using active travel modes because they cannot drive or do not have access to a vehicle may be able to use CAVs, thus reducing the proportion of people using active travel modes. Conversely, the improvements in road safety resulting from CAVs may lead to more people cycling or walking.

On average, today's cars are parked about 95% of the time [31] leaving huge transportation resources unused for a large portion of the day. With this technology being gradually introduced, we will have less parking in the city as most vehicles will be on the move. Furthermore, CAVs will be able to drive very closely from each other, without impacting on safety since machines required less time to react when a hazard occurs compared to normal human drivers. This will increase roadway capacity and allow more houses to be built, more business to settle and more commercial centres to open. We believe that these vehicles will be electric and possibly able to also charge themselves with limited or no human intervention. They will be using very powerful batteries. TESLA is one of the leading car manufacturers invested in car battery technology. This potential disruptive technology will create greener mobility with less CO_2 emission.

6 Barriers to Connected Autonomous Vehicles Implementation

Although CAVs offer considerable benefits, applications and opportunities in transportation, it is undeniable that their implementation will also present huge challenges to governments, car manufacturers and other related industries around who will have to face and work together to overcome the challenges. The speed and the nature of CAVs mass adoption are far from guaranteed. This will depend largely on how the technology is introduced to the market, their cost, the transportation laws and regulations put in place to preserve the safety and privacy of their users. The following sections outline some of the barriers to CAVs implementation. CAVs and AVs technologies are being tested in several cities across the world, global agreed standards and regulations are required and for the UK government to express their full confidence on the technology, the following challenges must be addressed.

6.1 Consumer Acceptance

Since Norman Bel Geddes envisioned cars able to drive themselves without human intervention in the 1939 World's Fair General Motors exhibit Futurama, AV technologies has significantly improved. Connected Autonomous Vehicles (CAVs) will be here much sooner than most people expect and will lead to major changes to transportation, our cities and society. The car manufacturing industry used to be the area of mechanical engineering. With advances in electronics, robotic and computer science, software companies such as Google are somehow leading the race. More than 50% of innovations in vehicles today are electronic. In the early 2000s, several universities took part in the Defense Advanced Research Projects Agency (DARPA) challenges in (2004, 2005 and 2007), most car manufacturers (Mercedes, BMW and Tesla) and some other technology companies (Google, UBER) are actively developing and testing AVs but there are several barriers to the introduction of this disruptive technology. At first, many consumers may be reluctant to put their lives in the hands of a robot. Recent studies and surveys have shown a split in opinion on whether people would like the autonomous capability to be available in their vehicles or not. Therefore, mass acceptance of this technology could take a long time. This could be the case particularly if there are accidents involving even semi-autonomous vehicles early in the adoption phase, whether it was the fault of the autonomous system or not [32, 33]. The transition from humans as drivers to humans as mere passengers in a car that drives itself is a major one. People generally have emotional connections with their vehicles. Therefore, are drivers willing to give up direct control over their vehicle and under what conditions? If automation of vehicles is not accepted by the users and users refrain from using the technology, the impact of automation on traffic flow efficiency, traffic safety and energy efficiency is mitigated. It is, however, not yet clear to what extent users accept automation and what the determinants of consumer acceptance of automation are [34]. Further, societal acceptance is pending with issues like safety, trust, security, privacy concerns, etc. Therefore, mass acceptance of this technology could take a long time.

6.2 Vehicle Costs

Reference [35] highlights that the cost of most autonomous car technologies applications for military and civilians is about $100,000. This is almost inaccessible for most people in the UK. Today, the high-end automotive Lidar systems mounted on the roof of these cars is estimated to about $75,000. The hope is that with mass production and notions related to Moore's Law may also apply here to allow the prices of this technology to come closer to the conventional vehicles' prices. J.D. Power and Associates' survey [36] found that 37% of persons would "definitely" or "probably" purchase a vehicle equipped with autonomous driving capabilities in their next vehicle. Nevertheless, costs remain high and is, therefore, a key implementation challenge, due to the current unaffordability of even some of the more basic technologies.

6.3 Legislation Liability and Litigation

In large cities, national and local authorities and law enforcement agencies will have to act swiftly in developing laws that allow cars to drive themselves on the streets without human intervention. Current legal systems have provision to deal with problems related to manufacturer defects. However, a framework for determining liability in a situation where an accident occurs while the vehicle is handing full control over to the human driver of the semi-automated technology. In this situation, there should be more clarity in the application of current civil and criminal law to shed a light on how to deal with the problem. When AVs and CAVs become certified for safe operation by the government, the regulatory bodies and other agencies responsible are to check that new technologies are of low risk, should new insurance and litigation issues arise. Such as the persuasion of insurance companies that they will work properly in all driving environments. The reality is that even with near-perfect autonomous driving, there may be instances where accidents are inevitable. Amongst the potential implications of this, people who otherwise are not able/allowed to drive could "get behind the wheel" of AVs or CAVs, and cars could technically drive from one place to another with no occupants. If there is an accident involving an autonomous vehicle, who is liable for the consequences as the driver is still behind the wheel? because the driver is still behind the wheel and therefore ultimately liable for the safety of the vehicle. But even this point may be intensely debatable [32, 33, 37]. It appears that with possible low number of accidents in fully connected smart cities, the insurance market will be disrupted. In general, most industries will have to re-invent themselves, change their business models or disappear altogether.

6.4 Social and Ethical Issues

Autonomous cars raise several kinds of ethical issues

(a) Is it possible to configure and programme a CAV to react to every single imaginable situation on the road? For example, not obeying traffic light signals or speed limits when driving someone in an emergency to the hospital (A&E) or dangerous driving in order to escape from a life-threatening circumstance.
(b) Although it is certain that CAVs will bring substantial social and economic benefits to cities around the world, several industries will be disrupted, and many people will lose their jobs and surely must change career.
(c) If an animal such as a deer jumps in front of the vehicle from nowhere, does the CAV hit the animal or run off the road? How do actions change if, instead of a deer, there is another car, or a pedestrian, a cyclist, a motorcyclist or even a heavy-duty truck? How does the algorithm developed in these vehicles react in those situations? With a split second for decision-making, human drivers typically are not held at fault when responding to circumstances beyond their control, regardless of whether their decision was the best at the time. In contrast, CAVs have sensors, visual interpretation software, and carefully designed programmes that enable them to potentially make more informed decisions. In a court of law, CAVs behaviour in some scenario may be questioned even if they are theoretically not "at fault". Other ethical question may arise concerning the algorithm in these technologies. How do they make their decisions on who to protect most or kill in a binary situation, for example who should be protected between five adults and a kid crossing the road or between a disable person and an elderly pedestrian? Should the vehicle owners be allowed to adjust such settings?

6.5 Cybersecurity, Data Security and Privacy Concerns

The idea that a car will be connected to the internet and able to drive itself without any human input raises several cybersecurity, data security, privacy, certification and licensing concerns as these vehicles may be subject to attack by criminals or terrorists and use for malicious purposes. These vehicles are just like computers on the road, therefore hackers may be able to take over control of the car either to kidnap someone or a group of people remotely, purposely create an accident, terrorists may be able to guide the stolen vehicle into a crowded area to kill people or load cars with explosives as a car bomb. Gang dealers may be able to deliver drugs or weapons including firearms to remote locations without being caught. It is clear that conventional cars are being used for some of the crimes listed above, but it is also obvious that with CAVs they will be achieved a lot more easily [29]. All countries face the same dilemma of how to fight cybercrime and how to effectively promote security to their citizens and organisations. Cybercrime, unlike traditional crime which is committed in one geographic location, is committed online and it is often not clearly linked to any geographic location [38]. Large scale cybersecurity attacks by hostile nations, disgruntled employees, terrorist organisations can be mounted on the whole city transportation system, disrupting traffic and creating collisions and all kind of accidents. For example, a computer virus could be designed to first infect virtually the entire UK CAV fleet as a dormant programme and later become active and create all kind of disaster on the road. Therefore, a coordinated global response to the problem of cybercrime is required. According to [39], the vice president of software security firm Vínsula, current cyberattacks are generally acts of espionage; most attackers gain unauthorised access to systems to gather information about their opponents rather than actual sabotage. Disrupting the vehicle electronic systems and sensors will require a more complex form of attack than the one used for data gathering which is generally harder. Regardless, the threat is real, and a security breach could have lasting repercussions [40]. Therefore, CAVs manufacturers, transportation policymakers and governments around the world should set security measures to handle these types of concerns. As CAV become mainstream and adopted around the world, privacy concerns will raise several questions: Who should own or control the vehicle's data? What types of data will be stored? With whom will these data sets be shared? In what ways will such data be made available? And, for what ends will they be used? In the UK, particularly in London, there are literally thousands of cameras watching us, some call London a "big brother state". From the moment you leave your home to the moment you get to work or school, you have an average of 300 cameras recording your movements. Our smart phones are equipped with location services. The reality is that privacy is almost a myth and these concepts are more likely to be transferred to CAV application. Someone involved in a car crash may not want his vehicles data to be shared with third parties, particularly if the person is at fault. Law enforcement could also benefit from such data. Risks such as losing privacy and/or integrity in the public cloud may prevent many decision makers to authorise the implementation of digital services using cloud computing in a smart city [41]. In this situation, sharing traveller data may be balanced with privacy concerns.

6.6 Infrastructure

To get the full benefits of CAVs, new road and communication infrastructure will be required. Since CAVs are not yet completely deployed, it may be difficult to guess every single road equipment that would be required. Although AVs generally require less bespoke road infrastructure, CAV must interact with their environment, communicating with other vehicles and roadside base stations in a Car2X context, they will be partially dependant on road infrastructures such as GPS mapping, road marking and strong telecom networks. The Transport Systems Catapult said that "infrastructure that is being imagined, designed, and built now, needs to have capability for future compatibility and functionality built-in from the get-go" [29]. This is certainly true because they are very expensive to maintain and upgrade.

7 Conclusion and Future Research

In this article, CAVs was explored as a potential solution to several issues faced by large cities around the world including excessive traffic jams, road accidents, CO_2 emissions and public health deterioration. It is undeniable that transportation has a massive impact on social welfare, urban sustainability. It can influence the growth of digital economies in large cities and CAVs offer potentially transformative benefits that can alleviate some of these concerns and lead the way to a greater level of sustainability. Transportation has immense implications for social welfare, economic development, and environmental sustainability. Congestion, environmental degradation, social inequity, and public health issues are problems that sustainable transport policies urgently need to resolve. Some of the challenges mainly related to security, privacy, cybersecurity, ethical, legal and infrastructure have also been explored. Amidst the growth of ICTs and the sharing economy, the protection of personal data and the security of communication networks are vital to ensure society capitalises on the gains from increased connectivity. This study serves to inform policymakers, scholars, and various stakeholders in the automotive industry of privacy and cybersecurity challenges of CAVs for achieving smart and sustainable cities. There are several smart city projects all over the world. These projects must be multidimensional and integrate the different action fields of the city, interacting with human and social capital. Technological solutions must be understood as the tool to achieve the smart city goals and to tackle the challenges these cities will face. The main objectives of these Smart City projects must be to solve urban problems in an efficient way to improve the sustainability of the city and quality of life of its inhabitants. Furthermore, governments around the world should have strategies to deal with privacy and cybersecurity concerns. Possible future research on Autonomous vehicles and connected autonomous vehicles could be to develop a model to measure people behavioural intention to use AVs and CAVs.

References

1. Batty, M., Axhausen, K.W., Giannotti, F.: Smart cities of the future. Eur. Phys. J. Spec. Top. **214**, 481–518 (2012)
2. Harrison, C., Eckman, B., Hamilton, R., Hartswick, P., Kalagnanam, J., Paraszczak, J., Williams, P.: IBM J. Res. Dev. **54**, 1 (2010)
3. United Nations: Sustainable development goals, United Nations: New York, NY, USA (2015)

4. Lim, H.S.M., Taeihagh, A.: Autonomous vehicles for smart and sustainable cities: an in-depth exploration of privacy and cybersecurity implications. Energies **11**(5), 1062 (2018). https://doi.org/10.3390/en11051062
5. Park, J.: Sensemaking/what will autonomous vehicles mean for sustainability? (2018). https://thefuturescentre.org/articles/11010/what-will-autonomous-vehicles-mean-sustainability. Accessed 10 July 2018
6. The telegraph: autonomous cars in smart cities (2017). https://www.telegraph.co.uk/business/risk-insights/autonomous-cars-in-smart-cities/. Accessed 10 July 2018
7. Lubell, S.: Here's how self-driving cars will transform your city. Wired (2016). https://www.wired.com/2016/10/heres-self-driving-cars-will-transform-city/. Accessed 10 July 2018
8. Parkin, J., Clark, B., Clayton, W., Ricci, M., Parkhurst, G.: Autonomous vehicle interactions in the urban street environment: a research agenda. Proc. Inst. Civ. Eng. Munic. Eng. **171**, 15–25 (2018)
9. Schaffers, H., Komninos, N., Pallot, M., Trousse, B., Nilsson, M., Oliveira, A.: Smart cities and the future internet: towards cooperation frameworks for open innovation. In: Future Internet Assembly, pp. 431–446 (2011)
10. McLaren, D., Agyeman, J.: Sharing cities: a case for truly smart and sustainable cities. MIT Press (2015). https://mitpress.mit.edu/books/sharing-cities. Accessed 10 July 2018. ISBN 9780262029728
11. Monzon, A.: Smart cities concept and challenges bases for the assessment of smart city projects. In: IEEE Smart Cities and Green ICT Systems (SMARTGREENS), pp. 1–11 (2015)
12. Dameri, R.P.: Comparing smart and digital city: initiatives and strategies in Amsterdam and Genoa. Are they digital and/or smart? In: Smart City. How to Create Public and Economic Value with High Technology in Urban Space, pp. 45–88. Springer, Heidelberg (2014)
13. Hall, P.: Creative cities and economic development. Urban Stud. **37**, 633–649 (2000)
14. Benevolo, C., Dameri, R.P., D'Auria, B.: Smart mobility in smart city. In: Empowering Organizations, pp. 13–28. Springer (2016)
15. Lawrence, F., Kavage, S., Litman, T.: Promoting public health through smart growth: building healthier communities through transportation and land use policies and practices. APA (2006)
16. Bencardino, M., Greco, I.: Smart communities. Social innovation at the service of the smart cities. Tema. J. Land Use Mob. Environ. (2014)
17. Maddox, J., Sweatman, P., Sayer, J.: Intelligent vehicles? Infrastructure to address transportation problems–a strategic approach. In: 24th International Technical Conference on The Enhanced Safety of Vehicles (ESV) (2015)
18. Fistola, R.: Gestione innovative della mobilità urbana: car sharing e ICT. Tema. J. Land Use Mob. Environ. **0**, 51–58 (2007)
19. CISCO: the internet of things, infographic (2011). http://blogs.cisco.com/news/the-internet-of-things-infographic. Accessed 28 July 2018
20. Eurostat: passenger mobility in Europe. European Commission (2007)
21. Eurostat: energy, transport and environment indicators. European Commission (2011)
22. Farsi, M., Hosseinian-Far, A., Daneshkhah, A., Sedighi, T.: Mathematical and computational modelling frameworks for integrated sustainability assessment (ISA). In: Hosseinian-Far, A., Ramachandran, M., Sarwar, D. (eds.) Strategyc Engineering for Cloud Computing and Big Data Analytics. Springer, Cham (2017)
23. Yim, S., Barrett S.: Public health impacts of combustion emissions in the United Kingdom. Dep. Aeronaut. Astronaut. Mass. Inst. Technol. Camb. US (2012)
24. US environmental protection agency (EPA): car pollution effects (2012)
25. European research cluster on the internet of things–IERC: the internet of things 2012-New Horizons, Cluster Book (2012)
26. Ross, A.K., Serle, J., Wills, T.: Tracking drone strikes in Afghanistan: a scope study. The bureau of investigative journalism (2014). https://v1.thebureauinvestigates.com/wp-content/uploads/2014/07/TBIJ-Afghanistan-Report.pdf. Accessed 25 Sept 2018
27. Slakey, F.: Robots versus humans: unmanned spacecraft are exploring the solar systems more cheaply and effectively than astronauts are (1999). http://web.mit.edu/writing/2009/Combined%20Make-up%20Readings.pdf. Accessed 25 Sept 2018

28. NASA: unmanned spacecraft for surveying earth's resources (2018). https://ntrs.nasa.gov/archive/nasa/casi.ntrs.nasa.gov/19720015637.pdf. Accessed 25 Sept 2018
29. STSC house of lords: connected and autonomous vehicles: the future? House of lords science and technology select committee (2017)
30. ATKINS: connected & autonomous vehicles, introducing the future of mobility (2016). http://www.atkinsglobal.co.uk/~/media/Files/A/Atkins-Corporate/uk-and-europe/uk-thought-leadership/reports/CAV_A4_080216.pdf. Accessed 14 July 2018
31. Morris, Z., D.: Today's cars are parked 95% of the time, furtune (2016). http://fortune.com/2016/03/13/cars-parked-95-percent-of-time/. Accessed 18 Aug 2018
32. SMMT: the society of motor manufacturers and traders motor industry facts 2013, s.l.: SMMT driving the motor industry (2013)
33. KPMG: connected and autonomous vehicles-the economic opportunity, s.l.: SMMT driving the motor industry (2015)
34. Hoogendoorn, R., et al.: Towards safe and efficient driving through vehicle automation: the dutch automated vehicle initiative (2013)
35. Dellenback, S.: Director, intelligent systems department, automation and data systems division, southwest research institute. Communication by email, May 26 (2013)
36. J.D. Power and associates: 2012 US automotive emerging technology study (2012)
37. McChristian, L., Corbett, R.: Regulation issues related to autonomous vehicles'. J. Insur. Regul. Natl. Assoc. Insur. Comm. (2016)
38. Jahankhani, H., Al-Nemrat, A., Hosseinian-Far, A.: Cybercrime classification and characteristics. In: Bosco, F., Staniforth, A., Akhgar, B. (eds.) Cyber Crime and Cyber Terrorism Investigator's Handbook, Syngress (2014)
39. Hickey, J.: Vice President, Vínsula. Telephone interview, October 11 (2012)
40. Fagnant, D.J., Kockelman, K.: Preparing a nation for autonomous vehicles: opportunities, barriers and policy recommendations. Transp. Res. Part A Elsevier, 167–181 (2015)
41. Hosseinian-Far, A., Ramachandran, M., Slack, C.L.: Emerging trends in cloud computing, big data, fog computing, IoT and smart living. In: Dastbaz, M., Arabnia, H., Akhgar, B. (eds.) Technology for Smart Futures. Springer, Cham (2018)

A Digital Twin Model for Enhancing Performance Measurement in Assembly Lines

Christos I. Papanagnou

Abstract Dynamic manufacturing processes are characterized by a lack of coordination, complexity and sheer volumes of data. Digital transformation technologies offer the manufacturers the capability to better monitor and control both assets and production. This provides also an ever-improving ability to investigate new products and production concepts in the virtual world while optimizing future production with IoT-captured data from different devices and shop floor machine centres. In this study, a digital twin is presented for an assembly line, where IoT-captured data is fed back into the digital twin enabling manufacturers to interface, analyse and measure the performance in real-time of a manufacturing process. The digital twin concept is then applied to an assembly production plan found in the automotive industry, where actual data is considered to analyse how the digital duplicate can be used to review activities and improve productivity within all production shifts.

Keywords Digital twins · Performance measurement · Assembly lines · Automotive industry

1 Introduction

The distribution of products with shortened life cycles, as well as the continuously increasing customer expectations have led manufacturing companies to invest more in technology and product augmentation [1, 25]. Production managers put efforts to reduce production costs significantly while maintaining excellent product quality and high levels of customer services. The globalization of markets together with the elimination of import trade duties and restrictions has also forced manufacturers to look for ways to improve their competitive positions by focusing on Research & Development (R&D) [5]. Many companies contemplate that significant (especially long-term) savings can be achieved by managing their manufacturing supply chain more effectively in the midst of investing in technology [24, 32].

C. I. Papanagnou (✉)
Salford Business School, University of Salford, Manchester M5 4WT, UK
e-mail: c.papanagnou@salford.ac.uk

M. Farsi et al. (eds.), *Digital Twin Technologies and Smart Cities*,
Internet of Things, https://doi.org/10.1007/978-3-030-18732-3_4

Manufacturing systems should be not only designed and operated for high reliability and throughput but also to have the capability to integrate the shop floor with other departments and sections of the business environment. Production planning in a dynamic business environment should have the capacity of dealing with uncertainty in line with satisfying customer delivery time, low cost and high quality. Therefore, manufacturing plants require a mechanism that monitors production process flow and normalization in the case of production disruption [8].

Manufacturing plants are often characterized by complexity and very often managers experience difficulties to perform in-depth data analysis and decision-making. In most cases, efforts are made to evaluate production data and elaborate results by improving the quality of products while reducing manufacturing costs [23]. Production planning in manufacturing involves in most cases the synchronization with the downstream demand and thereby has a strong impact in warehouses of both manufacturers and other supply chain participants [19]. By the manipulation of the production line, it is hoped that new knowledge about the production process can be obtained without the inconvenience or cost of manipulating the real process itself. Therefore, it becomes indispensable to understand production systems' behaviour and the parameters that affect the performance of production lines [11, 27].

In the past years, manufacturing companies have been able to reduce waste and volatility in their production processes and dramatically improve product quality and yield by applying lean techniques [1]. However, in certain processing environments, extreme swings in variability are still a fact of life, while the complexity of manufacturing systems complicates planning and scheduling for managers and operators. Often, the continuity of the production process is at risk due to the inadequate planning and control in the production systems [17]. Standstills in the manufacturing line—in the event of failure, commissioning, reconfiguration, adaptation or breakdown—apart from the disruption that cause can be very detrimental to the manufacturing company.

Production and manufacturing systems are characterized by a number of different performance measures including flexibility, resource and output measurement [5]. The goals of each of these three measures are different, and therefore, at least one individual goal type that corresponds with the organizations' strategic goals from the three listed measures must be present in a supply chain performance measurement system. According to Simpson et al. [27], a flexibility level system reacts to uncertainty issues when performance measurement is carried out at each node of an extended enterprise. Firms should have key performance indicators in the areas of cost, time, innovation, quality and precision corresponding with the mission and strategy according to stakeholders' perception.

Output measurements are often associated with throughput and average up-times. Thus, most of the manufacturers in order to increase the productivity tend to minimize the unavailability of the lines. As a result, reconfiguration of the manufacturing systems occurs (erroneously) only when essential work has to be done although upgrades of the system are desirable to increase quality, increase throughput or reduce energy consumption. Assessing real-time manufacturing environment involves understanding the dynamics affecting the performance. In a manufacturing environment, a precise performance measurement of supply chain activities is based on quantitative measures which are the utilization of resources and cost, quality and manufacturing flexibility. The related data to the quantitative key performance indicators (KPI) can be retrieved from annual and financial reports and company's management opinion. Very often, manufacturers invest in new machinery and online optimization after commissioning and installation. In high volume manufacturing and especially in automotive manufacturing, efficiency and good performance heavily depend on reliable and highly available manufacturing and automation systems.

Furthermore, existing literature suggests that essential knowledge such as information and advanced technology can enhance supply chain performance [5, 16]. However, there is a limited literature on how this knowledge can be applied in the manufacturing supply chain. Most of the studies in regard to measuring performance management of a manufacturing plant in the automotive industry are restricted in limited locations such as Brazil, India and Australia. This constitutes a limitation per se because the conclusions can be hardly generalized to other countries [14]. Therefore, it is vital to investigate how knowledge contributes to the automotive industry in another country as performance is subject to location, plant's setting and size among other factors [4].

1.1 Digital Twins in Manufacturing Context

In the last 5 years, technology and knowledge transfer within an industrial organization or between manufacturing plants have been reinforced by digital transformation and Internet of Things. The Fourth industrial revolution has started to reshape many organizations while digitalization has enabled companies to transform operational effectiveness, improve safety and increase production. However, as both complexity and uncertainty are always present in production lines, industrial organizations should make a further step beyond digitalization and consider a more granular virtual model approach to monitoring, diagnosing and correcting process flaws. This model approach constitutes a form of a digital twin.

Digital twins (DT) have been introduced initially as virtual clones to physical products, in order to improve geometry assurance in early product design phases or to observe and study certain aspects of the products without having to interfere or taking the product out of service [27, 31]. Tuegel et al. [29], propose a DT for predicting the life of aircraft structure and assuring its structural integrity while the system dynamics of a product were reinforced by DT for better interpretation of customers' needs [29].

The interest in digital twin technologies is rising as the concept of a smart digital factory and sensor-driven operations have gained the attention of many manufacturing companies [2]. DT enable autonomous objects to imitate the current state of processes and their own behaviour. Also, they can be used as a flexible data-centric communication middleware to develop a reliable advanced driver assistance system in autonomous systems such as self-driving cars [33]. In recent years, there is a focus on digital twin-driven manufacturing cyber-physical system (MCPS) for parallel controlling and simulation of shop floor processes [7, 18]. Thanks to the historical production data, manufacturers can apply computational methods to create a digital model of the manufacturing process whereas the use of real-time data from sensors may reduce waste, maximize throughput and conduct innovations. Alam and Saddik [3], introduced a digital twin architecture reference model to describe the properties of a cloud-based cyber-physical system (CPS) [3]. Modelling and simulation with the aid of DT may offer recommendations, support design tasks or validation of system properties [6].

This work suggests a DT platform, which replicates a complex manufacturing system and predicts future intervention requirements by supporting "ad-hoc" data analytics to maximize the performance of the factory. Apart from the palpable benefits to the manufacturing process, the proposed DT model coupled with the Internet of Things, big data analytics and cloud technologies can be used to drive growth in manufacturing and to open up new MCPS-based business models. Software companies develop DT technology that further builds out IoT capabilities in their enterprise asset management portfolio by allowing customers to leverage IoT data in creating a virtual model of an asset.

2 The Digital Twin Modelling Platform

Manufacturing systems require deeper analysis of various data from machine centres and processes. Although manufacturing companies take advantage of state-of-the-art modelling techniques and advanced systems, increasing complexity due to the large data arrival can be only addressed using appropriate distributed, interoperable, and high-performance ICT solutions. For that reason, DT technology, which is applied in dynamic manufacturing processes, should self-optimize, by capturing data from production and potentially, ambient data from various sensors, as well as data from operators and managers involved in the production process. The data feeds back into the DT, creating a closed loop that enables manufacturers to interface with an actual plant as if it was an Internet-based software.

An overview of the proposed DT platform within a manufacturing environment is depicted in Fig. 1. As it can be inferred, prediction techniques derived via the DT platform are able to forecast the ever-changing needs of plant facilities and to offer the potential of creating new markets. DT platform adopts and leverage (symbiotic) simulation techniques, and thus, it interacts with the physical system in a mutually beneficial way. Also, the DT platform is highly adaptive, in that the DT platform not only performs what-if experiments that are used to control the physical system but also validates and responds to data from the physical system via actuators [12].

There are still manufacturing companies that fail to exploit valuable big datasets created in process planning while working with quality and environmental standards [20]. Shop floor data analytics are very important as it is expected that data in manufacturing environments will increase exponentially during the following years.

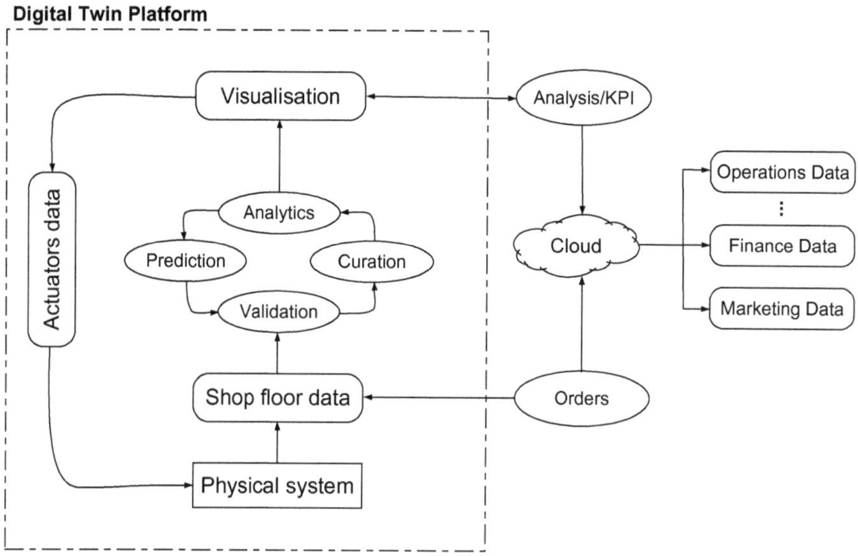

Fig. 1 Digital twin modelling platform

The DT platform can manipulate large amounts of shop floor data accrued from the physical system. The proposed DT platform may work as an enabler to manipulate big data and optimize the physical system by automatic control based on scenario testing of different variables in real time towards optimization. It can run in parallel with the manufacturing processes whilst constantly analysing, modelling and visualizing relevant data in real time. As a result, the DT platform may uncover and leverage data that is hidden or unappreciated so as to deliver information capable of transforming processes.

With the aid of big data analytics, the proposed DT platform provides also an integrated approach for advanced modelling, analysis, feedback and visualization techniques, which are helping manufacturing companies to eliminate waste and create value through the design and production of the products. Datasets are analysed against essential KPIs, while with the aid of a cloud this information is propagated to other departments and core activities of the company including marketing and finance. Thus, the DT platform depicted in Fig. 1 constitutes an inextricable "business as usual" strategic business module that offers further opportunities including product development acceleration, design methods that minimize production costs and a plethora of products that can be bought at better prices from customers. Moreover, DT platform harnesses consumer insights to reduce development costs through innovative approaches and customized products, which can mark the dawning of the Manufacturing as a Service innovation [9]. Things are already starting to move in that direction with companies such as Adidas with its innovative SpeedFactory facility, which produces semi-custom shoes and Nike, which just acquired computer vision firm Invertex [30].

With the ever-increasing complexity of manufacturing organization processes and business models, the challenge of linking high performance and quality with cost-effective productivity is always present. Most manufacturing companies in order to cope with downstream demand and new customer requirements follow the traditional three shifts scheme, which often increases in involvement in occupational injury for employees [28]. On the other hand, studies showed that night shifts operators manage their work–life balance without sacrificing productivity. From management's perspective, the performance for each shift is subject to throughput and shop floor data. Thus, a DT platform should also provide a transparent user interface of all relevant variables, such as throughput, and raw data derived from shop floor and especially machinery. In the next section, the purposed DT platform will be utilized to study the performance of a three-shift assembly setup in an automotive industry.

3 Case Study: The Adoption of Digital Twin Platform to Leverage the Performance of an Assembly Line

The motivation for this case study is to review the problems encountered within a complex manufacturing plant with respect to the execution of performance valuation within production management, as well as how these problems can be checked, controlled and enhanced within the performance management perspective of production. In accomplishing this motive, this study was centred on the assembly manufacture line of an automotive company situated in the United Kingdom. There are 56 machine centres (workstations) across the production line whereas parts are processed in a sequential manner.

The occurrence of machine breakdown is very common and uncertain in assembly production plants and repair times depend on the condition of the machine at each breakdown event. In this study, the overall time of machine breakdowns for all machine centres and for each 8 h shift in a single day is calculated. As the determination of the exact time of machine breakdown and the duration of the repairs is quite difficult, manufacturing companies fail in achieving an optimal plan of productivity or even due to order deliveries of end products under the given time horizon by the client. There are different reasons for machine breakdown varying from poor maintenance, machine deterioration and overruns to weather conditions and operator mistakes. However, there is a limited number of studies on the breakdown events. For example, changeovers may last longer than the work schedule due to operators' own volition or because machines' stoppage times can be elicited manually by operators for no certain reasons. Some studies suggest that preventive maintenance may lessen the likelihood of machine breakdown [10], however, the emphasis is given more on the reduction of maintenance cost rather than on operational performance and breakdown events [15]. Thus, a DT platform with the aid of an automated data collection system and actuators may restore machines' operation to the desired work levels.

This study opts for the acquisition of quantitative shop floor data that could assist in determining the impact of machines breakdown in performance by using the proposed DT platform depicted in Fig. 1. Furthermore, primary research helped to acquire important data that assisted in obtaining answers to the following questions identified in this study:

- Question 1: *What are the major difficulties and requirements with respect to affecting performance management within production management by utilizing a DT platform?*
- Question 2: *In what way can these difficulties and requirements be observed and addressed by deploying a DT platform in the performance management context?*

3.1 Validation and Curation of the Shop Floor Data

Different raw shop floor annual data (365 days) was gathered and aggregated. This data includes (a) the "flag" event data every time a breakdown arises accompanied by information of the date and the ID of the machine centre the breakdown occurred, and two timestamps signifying the start and end of the breakdown; (b) the temperature data from the machines centres, collected every second; and (c) the number and duration of jobs each machine performs in an hour. Due to the fact that all this information comes in raw format, validation helps to confirm the source of data in terms of origin and contain.

Curation of raw data from the shop floor is important in light of the fact that every company has its own methods of data collection. This data is extremely instrumental on the grounds that it helps in comprehending the company's machines behaviour under certain ambient conditions. The accumulation of raw semi-structured data poses the challenge of complexity in terms of analytics given the huge amount of data generated by the shop floor. Thus, this step provides a sustained and consistent form of systematic data curation and error prevention, which can eliminate bias in analysis and misinterpretation of machines behaviour. Validation and curation stage was processed by means of a numerical computing environment and proprietary programming languages.

3.2 Analytics and Prediction of the Shop Floor Data

After the data is validated and curated, analytics can identify certain machine activities and reveal important information (e.g., patterns) about the performance of each machine. In this section, the DT platform is utilized to exploit complex and large data to obtain trend analysis and prediction of breakdowns, which is otherwise an extremely complex time-consuming task and very often prone to errors. Most manufacturing organizations currently have systems which capture and store data from all business areas; however, they do not have a technology in place, which can provide trend analysis and prediction, but most importantly, they do not run a simulation in real time, which can optimize the physical processes.

The production plant's performance for each shift is measured by throughput rate (TR), which provides the number of finished products at a given time. In a traditional manufacturing, environment throughput is often subject to machines' breakdown times and production yield; which is expressed by the number of non-defective products divided with the total number of manufactured products [28]. By introducing a DT platform, data from compressing sensors measuring temperature changes in all machine centres is also considered to investigate whether temperature levels in addition to key shop floor data are associated with throughput rates. It should be also noted that data curation and validation of such data with the aid of visualization tools—provided a new set of variables, which constitute a "clean" format of manufacturing ambient data.

The aggregated breakdown times (BT) in seconds, production yield (PY) and average temperature values (TV) in °C for all 56 machine centres in a single day are used as independent variables in a two-step hierarchical multiple regression model, in order to investigate their effect on the throughput rates. Two-tailed correlations among the variables adopted in the analysis are shown in Table 1. In this study, Shift 1 is the night shift (22:00–06:00), Shift 2 the early shift (06:00–14:00) and Shift 3 is the late shift (14:00–22:00). It should be noted that almost all same types of shop floor data differ among the three shifts, with breakdown times between Shift 2 and Shift 3 the only exception. Also, the night shift has the lowest production yield, throughput rate and the longest breakdown times. This signifies that night shift has the worst performance, which may lead to long cycle times and an increase of control costs in factories [17].

To further understand whether key shop floor data is associated with productivity, hierarchical regression analyses were performed. Estimations based on a two-step hierarchical regression model for each shift are presented in Table 2. Initially, three different models were implemented to examine the linear relationship between the throughput rates with traditional key managerial data Model namely, A1, Model A2 and Model A3 for each shift, respectively. Then, the impact of the proposed digital twin platform was examined by adding the temperature levels for the 56 machine centres leading to Model B1, Model B2 and Model B3 for each shift, respectively. The initial findings suggest a significant direct association between throughput rates and shop floor data for Shift 1. The addition of temperature levels increases the regression model's R^2 from 0.229 to 0.636. Thus, the inclusion of temperature levels assisted by DT platform explains more than 63% of the variance in throughput rates (Model B1), while the production yield and breakdown times on their own explain 22.9% of the variance in throughput rates (Model A1).

In regard to Shift 2, the results in Table 2 suggest that for Model A2 the regression equation is not significant ($F = 2.058$). The addition of temperature levels in Model B2 improved significance ($F = 2.125$, $p < 0.1$), but only temperature levels variable is significant ($\beta = 2.125$, $p < 0.1$). The R^2 has slightly increased from Model A2 (0.077) to Model B2 (0.117), indicating a small contribution to the throughput rates

Table 1 Descriptive statistics and correlation among variables

	Mean	Min	Max	TR1	PY1	BT1	TV1	TR2	PY2	BT2	TV2	TR3	PY3	BT3
Throughput, shift 1 (TR1)	493.904	0.000	860.000	1.000										
Yield, shift 1 (PY1)	0.371	0.007	0.944	0.247*	1.000									
Breakdown, shift 1 (BT1)	99932.86	11820.394	379264.342	−0.263*	−0.009	1.000								
Temperature, shift 1 (TV1)	215.536	50.650	348.870	0.797†	0.133	−0.080	1.000							
Throughput, shift 2 (TR2)	590.231	0.000	870.000	0.126	0.002	−0.007	0.209	1.000						
Yield, shift 2 (PY2)	0.653	0.017	0.998	0.082	−0.194	−0.091	0.136	0.276$^{\diamond}$	1.000					
Breakdown, shift 2 (BT2)	97807.606	26940.073	459027.734	0.005	−0.260*	−0.054	−0.038	−0.458†	0.008	1.000				
Temperature, shift 2 (TV2)	269.282	45.670	383.730	0.168	0.218*	0.005	0.210	0.766†	0.153	−0.599†	1.000			
Throughput, shift 3 (TR3)	586.519	0.000	830.000	−0.116	−0.055	−0.249*	−0.247*	0.032	−0.105	−0.140	−0.052	1.000		
Yield, shift 3 (PY3)	0.715	0.075	0.997	0.135	0.039	−0.052	0.141	0.080	0.111	−0.072	0.122	0.150	1.000	
Breakdown, shift 3 (BT3)	98442.055	24619.447	330000.447	−0.091	−0.093	0.293$^{\diamond}$	0.012	0.178	0.219*	0.130	0.044	−0.679†	−0.279$^{\diamond}$	1.000
Temperature, shift 3, (TV3)	296.634	88.100	390.240	−0.211*	−0.197	0.107	−0.101	−0.071	−0.083	−0.141	−0.006	0.473†	−0.062	−0.278$^{\diamond}$

†significant at 0.01 level, $^{\diamond}$significant at 0.05 level, *significant at 0.1 level

Table 2 Results of Hierarchical Regression Analyses

	Dependent variable: Throughput rate					
	Shift 1		Shift 2		Shift 3	
Independent variables	Model A1	Model B1	Model A2	Model B2	Model A3	Model B3
Constant	589.077†	588.058†	473.062†	383.139†	797.894†	510.333†
Production yield	209.038*	170.030*	164.711	145.477*	−29.387	−1.056
Breakdown times	−0.002†	−0.001†	−0.005	0.000	−0.002†	−0.002†
Temperature levels		2.239†		0.417*		0.825†
Model *F*	7.262†	27.902†	2.058	2.125*	21.130†	18.903†
R^2	0.229	0.636	0.077	0.117	0.463	0.542

†significant at 0.01 level, ◇significant at 0.05 level, *significant at 0.1 level

for the Shift 2 by utilizing the DT platform. In contrast, the results for Shift 3 show that production yield is not significant to the throughput rates. This means that production yield numbers derived for Shift 3 do not provide a clear picture of the production performance. However, Model A3 and Model B3 suggest that the breakdown have a negative impact on throughput rates (β = -0.002, p <0.01. The inclusion of a DT platform in our analysis resulted to a highly significant model (Model B3) with $R^2 = 0.542$, as more than 50% of the variation in throughput rates can be explained by breakdown times and temperature levels. The analyses reveal that the insertion of a DT platform has a statistically significant positive relationship with the performance of the production plant. Note that temperature levels variable has the largest and highly significant coefficient in all six models, indicating that it is the most important factor, statistically, that could affect the performance by the means of throughput rates.

Results in Table 2 show that although a direct significant association exists between production's performance and breakdown times, the proposed DT platform helps to explain the influence of the machine centres' temperature values far more precisely and meaningfully. Note that, the results for Shift 3 indicate that problematic shifts in terms of throughput and can be linearly explained with the aid of breakdown times, production yield and temperature values. It is also clear that the proposed DT platform reinforces the initial results derived by the means of Model A1, as more than 63% of the variation in throughput rates can be explained by all independent variables in Model B1. Last but least, in order to ascertain the multicollinearity does not comprise an issue in shop floor data, the variance inflation factor (VIF) was derived for all models. The largest VIF score found was 5.67 (Model B3), which is below the maximum level of 10 that multicollinearity could cause unstable regression coefficients [22].

4 Managerial Implications

Even in complex manufacturing environments, continuous improvement and adoption of advanced technology to attain manufacturing excellence are often essential. Large investments at all of the products' life cycle from designing to re-engineering, involve decisions pertaining to technology and innovation management. The adoption of the proposed DT platform indicates that a manufacturer, as the recipient of knowledge, investor and decision maker, should actively seek how digital twins can be provisioned, realized and utilized within the manufacturing environment. As assembly lines consist of many workstations and may become very complex especially with large product variety [21], the beneficial role of a DT platform should be emphasized and encouraged.

As the vast majority of manufacturing companies rely on key performance indicators to assess the production performance versus operational costs and compliance (e.g., strict environmental laws and regulations) the interconnection of objects and processes via open virtual platforms becomes essential. The integration of computation with physical processes is not new as cyber-physical, sociotechnical systems and symbiotic simulation offer a plethora of advantages, however, manufacturers should be also able to monitor the behaviour of the physical asset in real life and embed technology seamlessly into core business processes. Thus, as throughput rates relate also with ambient data derived from machinery, DT platforms can enhance machine-to-machine communication to save energy and prevent machines precocious deterioration, and thus, minimize breakdown times and occurrences. The evaluation of data and information provides also the benefit of improving human–machine interaction (e.g., by introducing new technologies that promote the use of immersive data), which cultivates personnel' skills, performance and working conditions.

Managers from manufacturing companies should recognize that DT platforms are very important in order to simulate operations under different performances and predict key performance indicators with the actual behaviour of existing machinery. Visualization techniques can help also managers to understand whether a particular machine is reliable and switch from preventive maintenance to predictive maintenance. The findings of this study suggest that DT platforms give prominence to powerful simulation models that increase the accuracy and reliability of machines and controls within assembly production facilities. This is very important as the dependability on planned production sequence in assembly lines is very high [13].

5 Limitations and Future Directions

The proposed DT platform can assist manufacturing companies to become more competitive and generate the income that is required to cover labour costs and overheads of knowledge workers and to invest in environmentally friendly and worker-friendly factories. However, as sophisticated as a given DT technology might be initially, further studies should be undertaken to anticipate the sheer number of variables that can affect production, whether it is humidity, temperature, the intensity of use of a given machine and so on.

This study derives results from an automotive assembly line without investigating how the proposed DT platform may also reinforce machine-to-machine (M2M) communication by allowing cloud connectivity and integration resulting to speeding up manufacturing processes and optimal productivity. M2M technology helps to cope with the challenges of distributed devices and high data capacity by leveraging cloud infrastructures to enable assets spread across distributed manufacturing plants, which would be very helpful in complex assembly lines.

Last, the proposed DT platform should be tested in terms of supporting transmission status and exception information being processed on the fly by persistency engines and rendered on workstations through dedicated protocols. A further data analytics could reveal useful insights on how DT technology can provide state-of-the-art solutions for energy-efficient product life cycles and ECO usage for multi-modal visualization and interaction technologies. In addition, it should be investigated in the future how the proposed DT platform can facilitate better automation/self-assembly technologies for conventional workforce tasks (e.g., joining processes in a vehicle assembly line or mechanical fastening).

The proposed DT platform can be used as the bedrock to implement the next generation core virtual autonomous platform, which can be used by managers to gain insights into the manufacturing plant and strengthen the company's competitiveness. The exploration and analysis of diverse types of data can assist decision-making and add value to previously unexploited data streams, and thus, reduce the costs associated with data even involving personnel with less IT skills.

Manufacturing systems integration requires intelligent tools that will have the ability to monitor the plant floor assets, and predict the variation and performance loss. Digital twins can offer dynamic rescheduling of production and maintenance operations, and synchronize with other related business actions to achieve a complete integration between manufacturing systems and upper level enterprise application. It is expected that the proposed DT platform will reshape industrial production and service design in the name of future outcome-based value creation, mass customization and smarter cities, where citizens' demands and their consuming behaviour will become an integral part of the manufacturing process.

6 Conclusion

As many manufacturing companies still suffer from data transparency and shop floor complexity, this study proposes a sophisticated DT platform that can act as the beacon for manufacturing companies to put their big data insights into real-time action and not only map but also optimize their entire plant life cycle. An actual assembly line and real shop floor data have been used to associate and predict the production performance initially with breakdown times and production yield. The analysis throughout three different manufacturing shifts did not reveal initially safe deductions. Then, the results from the adoption of a DT platform and the inclusion of machines' temperature levels indicated that digital twins' technologies provide a better understanding on the relationships between shop floor data and production performance by the means of throughput levels.

References

1. Agus, A., Hajinoor, M.S.: Lean production supply chain management as driver towards enhancing product quality and business performance: case study of manufacturing companies in malaysia. Int. J. Qual. Reliab. Manag. **29**(1), 92–121 (2012)
2. Alaei, N., Rouvinen, A., Mikkola, A., Nikkilä, R.: Product processes based on digital twin. In: Commercial Vehicle Technology 2018, pp. 187–194. Springer (2018)
3. Alam, K.M., El Saddik, A.: C2ps: a digital twin architecture reference model for the cloud-based cyber-physical systems. IEEE Access **5**, 2050–2062 (2017)
4. Barnes, J., Morris, M.: Staying alive in the global automotive industry: what can developing economies learn from south africa about linking into global automotive value chains? Eur. J. Dev. Res. **20**(1), 31–55 (2008)
5. Beamon, B.M.: Supply chain design and analysis: models and methods. Int. J. Produc. Econ. **55**(3), 281294 (1998)
6. Boschert, S., Rosen, R.: Digital twin the simulation aspect. In: Mechatronic Futures, pp. 59–74. Springer (2016)
7. Brenner, B., Hummel, V.: Digital twin as enabler for an innovative digital shopfloor management system in the esb logistics learning factory at reutlingen-university. Procedia Manuf. **9**, 198–205 (2017)
8. Bunse, K., Vodicka, M., Schönsleben, P., Brülhart, M., Ernst, F.O.: Integrating energy efficiency performance in production management-gap analysis between industrial needs and scientific literature. J. Clean. Prod. **19**(6–7), 667–679 (2011)
9. Coullon, H., Noyé, J.: Reconsidering the Relationship Between Cloud Computing and Cloud Manufacturing, pp. 217–228. Springer International Publishing, Cham (2018)
10. Daneshkhah, A., Hosseinian-Far, A., Chatrabgoun, O.: Sustainable Maintenance Strategy Under Uncertainty in the Lifetime Distribution of Deteriorating Assets, pp. 29–50. Springer International Publishing, Cham (2017)
11. Ding, S.H., Kamaruddin, S.: Maintenance policy optimization–literature review and directions. Int. J. Adv. Manuf. Technol. **76**(5), 1263–1283 (2014)
12. Fujimoto, R., Lunceford, D., Page, E., Uhrmacher, A.M.: Grand challenges for modeling and simulation. Schloss Dagstuhl (350) (2002)
13. Golinska, P., Pawlewski, P., Fertsch, M.: Monitoring the operations management performance in automotive industry. In: Automation Congress, 2006. WAC'06. World, pp. 1–6. IEEE (2006)
14. Humphrey, J.: Globalization and supply chain networks: the auto industry in brazil and india. Global Netw. **3**(2), 121–141 (2003)
15. Ji, M., He, Y., Cheng, T.E.: Single-machine scheduling with periodic maintenance to minimize makespan. Comput. Oper. Res. **34**(6), 1764–1770 (2007)
16. Koperberg, S.: The information flows and supporting technology in the automotive supply chain: a suppliers focus. In: 6th Twente Student Conference on IT (2007)
17. Kren, L., Tyson, T.: Using cycle time to measure performance and control costs in focused factories. J. Cost Manag. **16**(6), 18–23 (2002)
18. Leng, J., Zhang, H., Yan, D., Liu, Q., Chen, X., Zhang, D.: Digital twin-driven manufacturing cyber-physical system for parallel controlling of smart workshop. J. Ambient Intell. Humaniz. Comput., 1–12 (2018)

19. Li, S., Yu, Z., Dong, M.: Construct the stable vendor managed inventory partnership through a profit-sharing approach. Int. J. Syst. Sci. **46**(2), 271–283 (2015)
20. Lundgren, M., Hedlind, M., Kjellberg, T.: Model driven manufacturing process design and managing quality. Procedia CIRP **50**, 299–304 (2016)
21. Make, M.R.A., Rashid, M.F.F.A., Razali, M.M.: A review of two-sided assembly line balancing problem. Int. J. Adv. Manuf. Technol. **89**(5–8), 1743–1763 (2017)
22. Marquaridt, D.W.: Generalized inverses, ridge regression, biased linear estimation, and non-linear estimation. Technometrics **12**(3), 591–612 (1970)
23. Norek, C.D., Pohlen, T.L.: Cost knowledge: a foundation for improving supply chain relationships. Int. J. Logist. Manag. **12**(1), 37–51 (2001)
24. Patel, P.C., Chrisman, J.J.: Risk abatement as a strategy for R&D investments in family firms. Strateg. Manag. J. **35**(4), 617–627 (2014)
25. Sanchez, A.M., Perez, M.P.: Supply chain flexibility and firm performance: a conceptual model and empirical study in the automotive industry. Int. J. Oper. Prod. Manag. **25**(7), 681–700 (2005)
26. Simpson, D., Power, D., Samson, D.: Greening the automotive supply chain: a relationship perspective. Int. J. Oper. Prod. Manag. **27**(1), 28–48 (2007)
27. Tao, F., Cheng, J., Qi, Q., Zhang, M., Zhang, H., Sui, F.: Digital twin-driven product design, manufacturing and service with big data. Int. J. Adv. Manuf. Technol. **94**(9–12), 3563–3576 (2018)
28. Trzcielinski, S., Karwowski, W.: Advances in the ergonomics in manufacturing: managing the enterprise of the future. In: AHFE Conference (2014)
29. Tuegel, E.J., Ingraffea, A.R., Eason, T.G., Spottswood, S.M.: Reengineering aircraft structural life prediction using a digital twin. Int. J. Aerosp. Eng. **2011** (2011)
30. Wang, L., Wang, X.V.: Latest Advancement in Cloud Technologies, pp. 3–31. Springer International Publishing, Cham (2018)
31. Wärmefjord, K., Söderberg, R., Lindkvist, L., Lindau, B., Carlson, J.S.: Inspection data to support a digital twin for geometry assurance. In: ASME 2017 International Mechanical Engineering Congress and Exposition, pp. V002T02A101–V002T02A101. American Society of Mechanical Engineers (2017)
32. Wen, K., Lin, Z.: The strategic evolution of foreign R&D investment in china. In: Proceedings of 2005 IEEE International Engineering Management Conference, 2005, vol. 1, pp. 119–123 (2005)
33. Yun, S., Park, J.H., Kim, W.T.: Data-centric middleware based digital twin platform for dependable cyber-physical systems. In: Ubiquitous and Future Networks (ICUFN), 2017 Ninth International Conference on, pp. 922–926. IEEE (2017)

Information Sharing in Sustainable Value Chain Network (SVCN)—The Perspective of Transportation in Cities

Luai Jraisat

Abstract The purpose of this paper is to explore the high-order themes to information sharing in sustainable value chain network (SVCN) with a focus on the applications of Internet of things (IoT) as an enabling innovative technology from the perception of the expert community. This research is an inductive study and adopts a multi-case study strategy in the context of smart transportation for freight flow in the UK. Twenty semi-structured interviews are conducted with experts in smart transportation projects. The phenomenon of information sharing is enabled by effective innovative technologies such as IoT. A conceptual framework is constructed by the themes of IoT applications and information sharing in SVCN.

Keywords Information sharing · Innovative technology · Internet of things · Sustainable value chain network

1 Introduction

A lack of information and understanding of transportation has a major role in smart cities. Improving such information is important for information sharing in sustainable value chain network (SVCN) of transportation in the smart cities. The phenomenon of information sharing is one of the key subjects to be enabled by effective information and communications technology (ICT) such as Internet of things (IoT) [1–3]. Optimizing the transportation activities with innovative ICT is considering smart solutions to support freight flow in urban areas due to the complexity of the processes taking place in transport systems and often conflicting expectations of stakeholders [4].

In fact, there are a great number of initiatives that are very close as for their objectives but they do not have a common basis like standards, conceptions, and strategies [5]. Since 2000, more than 40 different projects on smart transportation have been initiated in Europe [5, 6]. Nowadays, smart transportation of IoT includes not only a

L. Jraisat (✉)
Faculty of Business and Law, University of Northampton, Northampton, UK
e-mail: Luai.Jraisat@northampton.ac.uk

M. Farsi et al. (eds.), *Digital Twin Technologies and Smart Cities*,
Internet of Things, https://doi.org/10.1007/978-3-030-18732-3_5

great variety of information but also thousands of other systems using data to make intelligent transport-related decisions [7]. IoT technologies guarantee economic benefits as chain actors will be able to share valuable information and make decisions that are more reasonable. This is to reduce transportation time and transportation expenditures and the impact of transportation on the society and environment [8].

With visions from a multidisciplinary perspective, the IoT has become the common paradigm of modern ICT area by enabling innovative applications in nearly all sectors of the economy [8]. However, relatively little attention has been paid to the information sharing between actors enabled by IoT for smart transportation along the SVCN [3, 7]. Thus, this research aims to explore the high-order themes to information sharing in SVCN with a focus on the applications of IoT as a key enabling ICT innovative technology from the perception of the expert community.

This study will use the existing literature, as well as case studies to examine the IoT application and information sharing for smart transportation for freight flow in SVCN. This could be done by identifying the possible high-order themes to information sharing for smart transportation in SVCN with a focus on the applications of IoT as an enabling innovative technology. This can provide benefits in terms of sustainability chain performance [5, 7]. The study poses the following research questions:

RQ1. How can key themes of the applications of IoT be associated with information sharing in SVCN?
RQ2. How and why these key themes are effectively linked to information sharing in SVCN to improve value chain performance in practice?

This paper provides relevant views from the perspective of experts in smart transportation projects in the UK. The article starts with a theoretical background on SVCN, IoT and innovative technology, and information sharing. Next, the research methodology is outlined. Then, the key findings and discussion are presented. Lastly, conclusions are provided with managerial implications.

2 Theoretical Background

SVCN has been affected by the digital revolution where the actor's strategy is surrounded by this digital era that created a hub where everything will be connected to everything via the internet [8, 9]. SVCN is considered as the theoretical base of the information sharing phenomenon. SVCN should present a framework to researchers for solving information issues such as sharing, visibility, environment, sources, technology, and types [5, 8, 10]. The SVCN is an approach where delivery and transportation businesses are integrated with the growth of e-commerce in the EU. Hence, a roadmap for completing the market for transportation has identified the need for increased transparency and information to all chain network actors as a key objective for improving delivery operations and boosting e-commerce. European Commission Information Society [10] and Pang et al. [11] have identified that the revolution of

IoT is reshaping the modern chain networks with promising business prospects in order to create sustainable values for freight flow. The development in transportation is one of the factors to indicate the wellbeing of the country. Totally optimizing the logistics and transport activities with the support of advanced ICT in urban cities is considering the traffic environment, its congestion, safety, and energy savings within the framework of the market economy [12].

Historically, research on IoT has been linked to several themes, such as technology, collaboration, social networks, quality, costs, satisfaction, investment, system analysis, system control, and connectivity [5, 7, 11, 13]. In this scenario, the authors argued that it is important to identify a well-established approach to both IoT application and information sharing for smart transportation. Optimizing the transportation activities with innovative technologies is considering smart solutions to support freight flow in urban cities due to the complexity of the processes taking place in transport networks [4]. Thus, this will be leading to improved performance for a set of actors along the SVCN rather than a single actor [11, 13]. Although research does not ignore the importance of the IoT in SVCN, they do not focus fully on the application of the IoT in the mechanism of information sharing and especially in strategic issues such as challenges and benefits.

European Commission Information Society [10] has defined IoT as "Things having identities and virtual personalities operating in smart spaces using intelligent interfaces to connect and communicate within social, environmental, and user contexts". [14] defined IoT as a set of hardware, software, database, sensors, hub of database, and systems for the support of people. A foundational technology for the IoT is the radio-frequency identification (RFID) technology, which allows microchips to transmit the identification information to a reader through wireless communication [13]. In fact, the IoT is an enabling tool that leads physical objects to be alive and perform actions by connecting these objects, and then shares information for better decisions and improved performance within SVCN [8, 15]. To make these objects smart, digital technologies such as communication technology, Internet protocols, analytic systems, control system, and embedded devices can be applied to sensor networks [8, 16].

Researchers argue that the SVCN concept allows the focus on moving from a transaction to a relational perspective that considers the environment around actors or firms or objects [4]. This concept is a great interest in applying the perspective of the network to analyze information sharing within a value chain of smart transportation for freight flow in cities [3, 7]. Thus, when analyzing the association between IoT applications and information sharing, a lens should be highlighted on the cone-shaped concept map of the business network information ecological chain (BNIEC) illustrated by [13].

In SVCN, the first stage of value added is to create benefits and minimize challenges for the actors involving in smart transportation for freight flow [10, 11]. This focuses on various issues associated with information problems, improves information value, and enhances performance for all actors [13]. This adds value for type 1 of stakeholders such as citizens, drivers, public transportation managers, and local city administration [17].

The second stage is IoT components: information, information technology, information subjects, and information environment. This is directly connected with the concept of smart transportation in order to deal with three main conceptions: transportation analytic, transportation control, and vehicle connectivity. This brings value assessment and business technology application for type 2, which includes stakeholders such as data experts, database designers, transportation experts, traffic experts, logistic experts, communication engineers, network engineers, system designers, and sustainability experts [17].

The third stage is the links between different actors and objects in the transportation systems. Surrounding the focal actor, all actors share information as different logical roles. This is to develop interaction between data hub and transportation system for type 3 of stakeholders, namely data source providers, local services, data management, and communication technology. For example, the application of the road condition monitoring and alert systems is the most important of IoT transformation applications [2]. New distribution systems in cities are called to apply smart solutions to enhance transport for goods in cities in order to minimize the complexity of the urban transport systems and often conflicting expectations of the road users and other stakeholders of urban freight transport (city administrators, inhabitants, entrepreneurs, and shippers) [18]. The main idea of the concept of smart transportation and green mobility is to apply the principles of crowdsourcing and participatory sensing. This can be supported via various data sources from vehicles, sensors, data centers, infrastructure, smart phones, etc. in addition, smart transportation consists of key communication ways such as machine-to-machine (M2M) communications, which include vehicle-to-vehicle (V2V) and vehicle-to-infrastructure (V2I) communications for better links.

The fourth stage is the phenomenon of the present research, which is information sharing in SVCN. This stage illustrates how IoT applications facilitate information sharing based on hosting an aggregative information field. This is where all actors or objects "things" of the above types share their hub or database contents with the potential actors (e.g., customer, middleman, retailer, service provider and object) for increasing more benefits and decreasing existing challenges.

In fact, there are a great number of initiatives which are very close as for their objectives and tasks but they do not have a common basis like standards, conceptions, and strategies [5]. Reviewing the milestones have been reached in Europe, for example, the first research programmes for cooperative smart transportation date back to the 1980s; the European project PROMETHEUS (1987–1994) by using inter-vehicle communication in the 57 GHz frequency band [6]. By 2000, a new technology was initiated worldwide, triggered by the availability of GPS, embedded systems, and Wi-Fi. In Europe, more than 40 different projects on cooperative smart transportation have been initiated since 2000 [5, 6]: initial feasibility studies (i.e., FleetNet and NoW), technology state and standardization (i.e., SAFESPOT, GeoNet, SEVECOM, CoVeL, and COMeSafety), field operation tests on safety and traffic efficiency (i.e., DRIVE C2X, SIM-TD, SCORE@F, etc.), and cooperative automated driving (i.e., AutoNet2030 project). Actually, by means of information sharing among vehicles, as well as between vehicles and the roadside infrastructure, vehicles transform from autonomous systems into cooperative systems [5]. Nowadays, smart transportation associated with IoT is the largest and the most versatile group. It includes not only a great variety of information, road, navigating, and car systems but also insurance and control systems for a vehicle/driver (telematics) and thousands other systems using data to make "intelligent" transport-related decisions. IoT technologies guarantee enormous economic benefits as both carriers and transport users will be able to make more reasonable decisions to reduce passengers and cargo transportation time and to cut transportation expenditures and delays. In addition, "green" IoT applies technologies to reduce the impact of passengers and cargo transportation on the environment.

3 Methodology

This research is an inductive qualitative study and adopts a case study strategy. From a multidisciplinary perspective, a conceptual framework can be developed from both the existing literature and contextual field data [19]. The cases are projects in the context of transportation for freight flow in the UK. 20 Semi-structured interviews are conducted with experts in these projects. This research applies within case and cross-case analyses [20]. By defining the themes of the associations between IoT applications and information sharing in SVCN, it became possible to develop the framework. These projects are selected because they have smart ICT technologies (e.g., IoT) for transportation, and have focused on information sharing in their SVCN. Experts as key informants are chosen because they provide an overview of the IoT application, information sharing, and their projects as a whole. The aim is to gain a rich understanding of what are the applications of IoT in smart transportation, how far IoT enable information sharing between actors, and what the roles of information sharing in SVCN. The UK is one of the key countries which has initiatives in applying IoT to support sustainable development in sectors especially transportation in cities.

The sampling selection is based on the advanced research of the online directory of sustainable projects in the UK and it included projects that have been applied for smart transportation for freight flow in cities. This led to a list of 30 projects, which were then shortlisted to 10 projects based on three steps: satisfactory achievement records, positive email responses, and an initial phone interview. Then, two experts in each project were asked to identify a network of smart transportation for freight flow in order to form the unit of analysis as an SVCN. This is where 10 different projects (Case 1–10) of similar 10 SVCN (unit of analysis) of 2 different experts (subunit of analysis) are examined. Table 1 illustrates the selected projects and their details.

Each SVCN is formed of a set of stakeholders: type 1 of citizens, drivers, public transportation managers and local city administration; type 2 of data experts, database designers, transportation experts, traffic experts, logistic experts, communication engineers, network engineers, system designers and sustainability experts; type 3 of data source providers, local services, data management and communication

Table 1 Case study in the context of SVCN in the UK

Case	Interviewee	Unit of analysis	Project years	Project status	City
SVCN 1	Public transportation manager, local city administration	Type 1	2016–	Active	London
SVCN 2	Public transportation manager, local city administration	Type 1	2016–	Active	London
SVCN 3	Public transportation manager, local city administration	Type 1	2016–	Active	Bristol
SVCN 4	Data expert, sustainability expert	Type 2	2016–	Active	Birmingham
SVCN 5	Communication technology manager and service manager	Type 3	2017–	Active	Cambridge
SVCN 6	Communication expert, network expert	Type 4	2017–	Active	London
SVCN 7	Public transportation manager, local city administration	Type 1	2015–17	Inactive	London
SVCN 8	Data expert, sustainability expert	Type 2	2013–17	Inactive	Bristol
SVCN 9	communication technology manager and service manager	Type 3	2010–13	Inactive	Birmingham
SVCN 10	Communication expert, network expert	Type 4	2010–12	Inactive	Newcastle

Source The author's own work

technology; type 4 of all stakeholders. The interviews were conducted and recorded by the author in person, who was asked the same questions. The interviews were also transcribed and then sent to the experts for revisions. The approved interviews were used to develop case studies, which were analyzed through cross-case analyses [20].

4 Findings and Discussions

The intention of the present research is to contribute to the body of knowledge by providing new propositions for information sharing in SVCN with a focus on the applications of the Internet of things (IoT) as an enabling innovative technology from the perception of project experts.

At the cross-case level, to answer RQ1, key themes of the IoT applications that can be associated for information sharing in SVCN are explored. The exploratory case studies have indicated that the key themes should be categorized related to the four stages: stage 1—value added, stage 2—Linking IoT components to the concept of smart transportation in order to deal with IoT conceptions, stage 3—links between different actors and objects in the transportation system and stage 4—all actors or objects "things" of the above types share their hub or database contents with the potential actors. The research applies this cross-analysis to develop data exploration to enhance replication logic amongst the 10 cases (10 SVCNs), providing the views of 20 project experts. In Table 2, the cross-case matrix is to show the stages of IoT applications that smart transportation projects follow to create information sharing in SVCN with a focus on increasing benefits and decreasing challenges for better performance.

This analysis resulted in 14 first-order themes for IoT applications, which were then coded as 5 second-order themes that turned into 4 aggregate dimensions. These aggregate dimensions are associated with one overarching theme, information sharing for SVCN, in order to establish the theoretical association for the current study.

Researchers have proposed key findings to carry out development in SVCN for smart transportation in cities [4, 5, 8]. However, a wider body of knowledge about SVCN associated with IoT is needed to overcome overlapping concepts in order to generate consistent findings [5, 6, 13]. Thus, the intention of the current research is to contribute to the body of knowledge by providing a new conceptual framework for information sharing in SVCN attached to IoT as an innovative technology in smart transportation. The framework in Fig. 1 illustrates key themes effectively linked to information sharing in SVCN and thus, in order to improve value chain performance in practice. Amongst these, information sharing has become the central theme, which is formed by themes of IoT applications as antecedences for information sharing.

There has been an important interest in applying the concept of a sustainable network to understand complex interaction and applications within value chains [4, 5, 13, 21, 22]. The research proposes the conceptual framework that encompasses three key aspects of a sustainable network: Innovative technology (IoT application), information sharing, and value chain performance for smart transportation in cities.

Table 2 IoT applications across the 10 cases of smart transportation projects

Aggregate dimension	Second order theme: First order themes	Case									
		1	2	3	4	5	6	7	8	9	10
Stage 1	Benefits and challenges:										
	▪ identify information problems	X	X	X	X	X	X	X	X	X	X
	▪ improve information value	X		X			X	X			
	▪ enhance performance		X		X	X	X		X	X	X
Stage 2	IoT components:										
	▪ information	X	X	X	X	X	X	X	X	X	X
	▪ information technology	X	X	X	X	X	X	X	X	X	X
	▪ information subjects	X	X			X					X
	▪ information environment	X									X
	IoT main conceptions:										
	▪ transportation analytic	X		X	X		X		X		X
	▪ transportation control	X	X	X	X	X	X	X	X	X	X
	▪ vehicle connectivity	X						X			
Stage 3	Actor Interaction:										
	▪ data hub	X		X					X		
	▪ transportation system	X	X		X	X	X	X	X	X	X
Stage 4	Facilitating information sharing:										
	▪ increasing more benefits	X	X	X	X	X	X		X		X
	▪ decreasing existing challenges.	X	X		X	X	X	X	X	X	X

Source The author own work

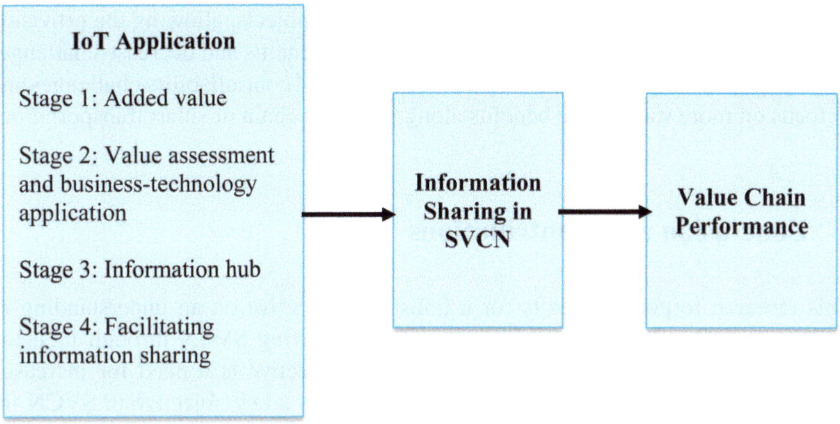

Fig. 1 A conceptual model for IoT applications and information sharing in SVCN. *Source* The author's own work

The present research provides brief explanations for each part of the proposed framework. First, the IoT application [13]. IoT as an innovative technology is proposed from the application used by various projects in smart transportation in the five cities undertaken in this study. The key findings highlight the importance of the four stages of IoT application to create information sharing in SVCN. The experts indicated that nine themes, namely identify information problems, enhance performance, identify information, information technology, transportation analytic, transportation control, transportation system, increasing more benefits, and decreasing existing challenges, are the highest important concepts and activities that should be included in IoT applications. The experts also indicated that five themes, namely improve information value, information subjects, information environment, vehicle connectivity, and data hub are the lowest important activities that can be included in IoT applications. The proposed framework indicates a set of recommendations for policymakers and projects' management.

In total, 14 themes linked to the four stages of IoT applications have an impact on creating information sharing in SVCN for smart transportation in cities. This, in turn, can bring improved value chain performance with a focus on sustainability aspects of economic, social, and environmental issues. The experts in all cases have illustrated the importance of these sustainability issues that have the potential to improve an efficient and effective transportation system for smart cities.

For example, key enabling innovative technologies including M2M, V2V, and V2I are IoT application technologies of communications for better links amongst actors. With multiple visions from various viewpoints, the IoT has become a key strategy in many smart cities [4, 5, 13]. IoT offers key benefits to various actors along the value chain including business to business and business to consumer, in addition to private and public sectors by enabling innovative applications. These applications provide a hub of information sharing for all actors based on combination of information technology, telecommunication, and objects, allowing the provision of valuable information on time. This can increase benefits and decrease challenges providing promising potentials to address visibility and controllability challenges and to focus on more sustainable benefits along the value chain of smart transportation.

5 Conclusion and Contributions

This research responds to calls for a holistic perspective on an understanding of how information sharing contributes towards improving SVCN through focusing on innovative technology [7, 23]. A holistic perspective is a need for increased transparency and shared information for all actors as a key objective in SVCN for improving smart transportation operations by IoT.

With multiple visions from different viewpoints, the IoT has become the common paradigm of modern ICT area [24]. It offers immense potential to consumers, companies, and public sectors by enabling innovative applications. This focus is attracting increasing attention from both policymakers and academics where prior research has suggested that this focus exhibits many unclear characteristics [25, 26]. There is a lack of how IoT applications can improve businesses in a sustainable way. Thus, this research aims to explore the high-order themes to information sharing in SVCN with a focus on the applications of IoT in transportation operations as a key enabling ICT technology from the perspective of the expert community along the value chain. In this research, a conceptual framework for information sharing in SVCN associated with IoT for transportation operations is then proposed.

References

1. Lindholm, M.: A sustainable perspective on urban freight transport: factors affecting local authorities in the planning procedures. Procedia Soc. Behav. Sci. **2**(3), 6205–6216 (2010)
2. Mirzabeiki, V.: An overview of freight intelligent transportation systems. Int. J. Logist. Syst. Manag. **14**(4), 473–489 (2013)
3. Andersson, P., Mattsson, L.-G.: Service innovations enabled by Internet of things. IMP J. **9**(1), 85–106 (2015)
4. Tachizawa, E.M., Alvarez-Gil, M.J., Montes-Sancho, M.J.: How "smart cities" will change supply chain management. Supply Chain Manag. Int. J. **20**(3), 237–248 (2015)
5. Vovk, Y.: Resource-efficient intelligent transportation systems as a basis for sustainable development. Overview of initiatives and strategies. J. Sustain. Dev. Transp. Logist. **1**(1), 6–10 (2016)
6. Festag, A.: Cooperative intelligent transport systems standards in europe. IEEE Commun. Mag. **52**(12), 166–172 (2014)
7. Uden, L., He, W.: How the internet of things can help knowledge management: a case study from the automotive domain. J. Knowl. Manag. **21**(1), 57–70 (2017)
8. Haddud, A., DeSouza, A., Khare, A., Lee, H.: Examining potential benefits and challenges associated with the Internet of Things integration in supply chains. J. Manuf. Technol. Manag. **28**(8), 1055–1085 (2017)

9. Chase, C.: The Digital Revolution: Crossing the digital divide is changing the Supply Chain Landscape—SAS Voices' (2016). https://blogs.sas.com/content/sascom/2016/04/19/crossing-the-digital-divide/. Accessed 04 Jan 2019
10. European Commission Information Society, 'Internet of Things Strategic Research Roadmap Antoine de Saint-Exupery' (2009)
11. Pang, Z., Chen, Q., Han, W., Zheng, L.: Value-centric design of the internet-of-things solution for food supply chain: value creation, sensor portfolio and information fusion. Inf. Syst. Front. **17**(2), 289–319 (2015)
12. Taniguchi, E., Thompson, R.G., Yamada, T., van Duin, R.: City logistics : network modeling and intelligent transport systems. Pergamon (2001)
13. Xu, X., He, W., Yin, P., Xu, X., Wang, Y., Zhang, H.: Business network information ecological chain. Internet Res. **26**(2), 446–459 (2016)
14. Wu, L., Yue, X., Jin, A., Yen, D.C.: Smart supply chain management: a review and implications for future research. Int. J. Logist. Manag. **27**(2), 395–417 (2016)
15. Gartner Press Release, 'Top 10 Internet of Things technologies for 2017 and 2018—Google Search' (2018). https://www.google.com/search?safe=active&rlz=1C1GCEU_enGB821GB821&ei=EnwvXLinGPy01fAPssqE-AQ&q=Top+10+Internet+of+Things+technologies+for+2017+and+2018&oq=Top+10+Internet+of+Things+technologies+for+2017+and+2018&gs_l=psy-ab.3..0j0i22i30.35895.35895..3643
16. Fang, C., Liu, X., Pardalos, P.M., Pei, J.: Optimization for a three-stage production system in the Internet of Things: procurement, production and product recovery, and acquisition. Int. J. Adv. Manuf. Technol. **83**(5–8), 689–710 (2016)
17. European Commission, 'Intelligent Transport Systems in action' (2011)
18. Małecki, K., Stanisław, I., Kijewska, K.: Influence of intelligent transportation systems on reduction of the environmental negative impact of urban freight transport based on Szczecin example. Procedia Soc. Behav. Sci. **151**, 215–229 (2014)
19. Eisenhardt, K.M.: Building theories from case study research. Acad. Manag. Rev. **14**(4), 532 (1989)
20. Miles, M.B., Huberman, A.M., Saldaña, J.: Qualitative data analysis: a methods sourcebook. Arizona State University, USA (1994)
21. Caragliu, A., Del Bo, C., Nijkamp, P.: Smart Cities in Europe. J. Urban Technol. **18**(2), 65–82 (2011)
22. Pilbeam, C., Alvarez, G., Wilson, H.: The governance of supply networks: a systematic literature review. Supply Chain Manag. An Int. J. **17**(4), 358–376 (2012)
23. Taniguchi, E., Thompson, R.G., Yamada, T.: Emerging techniques for enhancing the practical application of city logistics models. Procedia Soc. Behav. Sci. **39**, 3–18 (2012)
24. Atzori, L., Iera, A., Morabito, G.: The internet of things: a survey. Comput. Networks **54**(15), 2787–2805 (2010)
25. Porter, M.E., Millar, V.E.: How information gives you competitive advantage harvard business review. Harv. Bus. Rev. **63**(4), 149–160 (1985)
26. Browne, M., Gomez, M.: The impact on urban distribution operations of upstream supply chain constraints. Int. J. Phys. Distrib. Logist. Manag. **41**(9), 896–912 (2011)

Health Care in the Cyberspace: Medical Cyber-Physical System and Digital Twin Challenges

Jaime Ibarra Jimenez, Hamid Jahankhani and Stefan Kendzierskyj

Abstract Cyber-Physical Systems and Digital Twins are commonly used today in the industrial sector, and the healthcare sector is keen to implement these technological solutions to enhance their capabilities and offer better services for patient care provision. In fact, the adoption of Wireless Body Area Networks (WBAN) based on IoT along with cloud computing systems has led to the development of new methodologies to monitor and treat patients. However, the adoption of the new technologies comes with several challenges in terms of performance and security. Considering that, WBAN can be wearable or implanted under the skin, and the overall concept leads to several cybersecurity challenges that would require deeper investigation. This chapter presents an analysis of the impact that WBAN has on health care. It also provides some definitions of Medical Cyber-Physical Systems (MCPSs) and Digital Twins along with technological enablers such as cloud and IoT.

Keywords Digital twin · Medical Cyber-Physical system · Internet of things · Wireless body area networks · Biohacking · Personal health information · MCPS · WBAN · VM · Hypervisor

1 Introduction

Adoption of Cyber-Physical Systems (CPS) is gaining pace among most of industrial organisations and this trend is also being observed in the healthcare sector. *Medical Cyber-Physical Systems* (MCPS) is defined as critical, networked, distributed and context-aware systems of devices used in medicine. Therefore, MCPSs connect the physical and digital environment through network connectivity along with embedded

J. I. Jimenez (✉) · H. Jahankhani · S. Kendzierskyj
Northumbria University, London, London, Greater London, UK
e-mail: jaime.jimenez@northumbria.ac.uk

H. Jahankhani
e-mail: hamid.jahankhani@northumbria.ac.uk

S. Kendzierskyj
e-mail: stefan.kend@gmail.com

M. Farsi et al. (eds.), *Digital Twin Technologies and Smart Cities*, Internet of Things, https://doi.org/10.1007/978-3-030-18732-3_6

software. Likewise, Digital Twin represents the digitalisation of physical devices and artefacts. This technology has been used in industry, allowing it to simulate physical environments and specific machinery pieces in order to make decisions and assess risks in virtual environments prior to its implementation. This is the similar within the context of health care. Patients' body and physiognomy data are monitored on a real-time 24/7 basis with a view to provide more informed and real-time relevant healthcare responses.

Even though the Digital Twin is not implemented fully yet in medical setting, it is a topic that deserves further investigation, opening numerous challenges and research directions. It is an emerging engineering paradigm, which could allow healthcare data-driven practices such as delivering customised 3D printing prosthesis to be applied for surgical practices for instance. A Digital Twin in health care would take the concept of MCPS to a higher level. This can be viewed as a step closer to implementation of real smart city. The term 'Digital Twin' connects the physical and digital world, allowing users to visualise information of interest, on computers, mobile devices or even in holographic projections. The connectivity in Digital Twins, irrespective of the application context is enabled using sensory systems such as the ones used in embedded systems and Internet of Things (IoT). The concept of Digital Twin has been applied by NASA for the development and monitoring of aerospace vehicles.

This paper comprehensively and critically analyses the challenges that MCPSs and Digital Twins generate with a focused view on performance and security. Society is facing an era of an interconnected world, a 'Cyberspace', where devices, data, people, 'everything' are interconnected. Therefore, there is a need to conduct research on cutting-edge technologies prior to their design, configuration, implementation, monitoring and maintenance. The rest of the paper is organised as follows: Sect. 2 provides a brief discussion on MCPSs and Digital Twins. Section 3 analyses the impacts of Wireless Body Area Networks (WBAN) in health care. Section 4 discusses the challenges of MCPSs and Digital Twins in terms of performance, security and privacy. Finally, Sect. 5 concludes this chapter and suggests further research.

2 Medical Cyber-Physical Systems and Digital Twins

2.1 Medical Cyber-Physical Systems

MCPSs can be defined as intelligent systems related to medical devices [1] regardless of where they are being used (within hospitals, clinics or via wearable devices (see Sect. 3 regarding WBAN)). MCPSs are interconnected in the cyberspace using different networking protocols, frameworks and standards and are being considered in some countries such as Australia and UK as part of their Critical National Infrastructure (CNI). In addition, they are processed and manipulated via embedded software applications and monitored by caregivers. CPSs, which have been implemented in

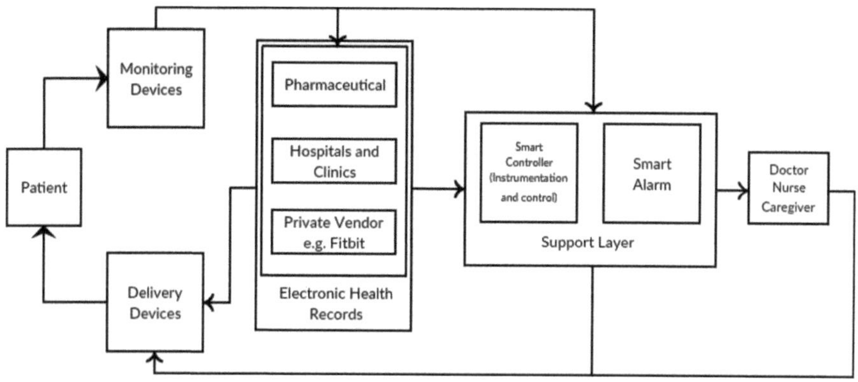

Fig. 1 Overview of a medical cyber-physical system, drawn from data provided in [1]

other sectors, utilised electronic sensors connected to Programmable Logic Controllers (PLCs) for sending physical information to customised applications, allowing operators to make decisions over their mechatronic infrastructures. The main difference in the medical CPS setting would be that sensors are allocated within medical devices connected to computer networks or used daily by patients through wearable devices (e.g., smartwatches and smartphones).

The values measured by sensors are triggered to transmit data in the following scenarios: (1) Within hospitals, communications can be triggered through Wireless Sensor Networks (WSN). (2) Using wearable devices, sensors send relevant information to applications installed in mobile devices through other wireless technologies (i.e., Bluetooth, ZigBee and radio frequency). Both cases offer a real-time monitoring while doctors can immediately evaluate any threats that can compromise a patient's health status [2]. The information can be available at Electronic Health Records (EHR) by accessing any administrative entity such as hospitals or clinics, pharmaceutical stores or private entities. These data are generally allocated in cloud computing systems and thanks to this technology, patients have faster access and better access rights to their health information compared to caregivers. One curious feature of MCPSs is the implementation of decision support devices formed by electronic and instrumented circuits which can trigger an alarm when an abnormal behaviour is observed with the patient health status. The support layer allows caregivers to make decisions in order to enhance the patient status [1]. In this architecture, the devices can be divided into two categories: monitoring devices used as sensors and delivery devices such as actuators which get modified according to the caregiver decisions (please see Fig. 1).

In addition, the research from Kocabas et al. [3] provides a general MCPS architecture divided into four layers, which are illustrated in Fig. 2.

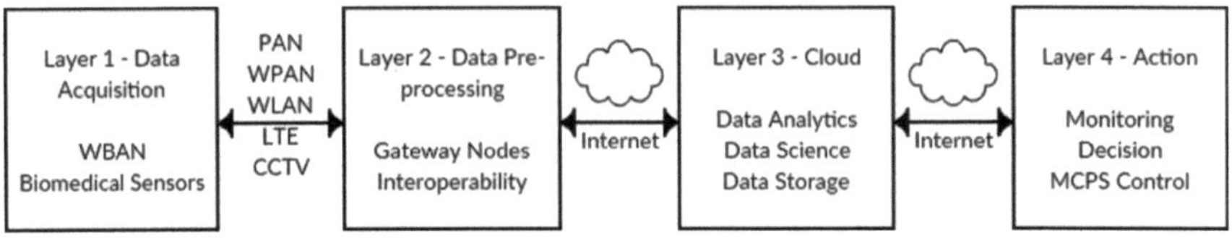

Fig. 2 Medical cyber-physical system architecture, drawn from data provided in [3]

2.1.1 Layer 1—Data Acquisition

Considered normally a Wireless Body Area Network (WBAN) using wireless protocols [4] in order to acquire interaction with the Internet using technologies such as ZigBee, WSNs, Bluetooth, Wireless Local Area Networks (WLAN), Wireless Personal Area Networks (WPAN), video surveillance systems and mobile networks [2, 4]. In the healthcare context, a WBAN uses biomedical sensors placed in different parts of the body and can be either wearable or implanted under the skin [5], allowing a real-time monitoring of the patient's health status such as blood pressure or body temperature, for instance.

2.1.2 Layer 2—Data Preprocessing

Currently, sensors have low computational power and are limited in hardware resources. Due to the high amounts of data gathered, it must be transmitted to a more sophisticated device prior to sending it throughout the Internet [2] in order to adapt the packet to traditional TCP/IP networks. Sensors require a gateway node (which acts as a concentrator) using a wireless communication. The concentrator allows these wearables or implanted devices to enhance their data concentration capabilities in order to transmit the information to cloud servers [6]. Likewise, the cloudlet has similar features compared to the concentrator; however, it is applied for more powerful devices, e.g., a smartphone. Typically, a cloudlet has a dedicated Internet connection and is configured from a dedicated computer as well [7, 8].

2.1.3 Layer 3—Cloud System

The cloud allows users such as doctors, nurses or patients to access Personal Health Information (PHI). It can be used to perform data analytics to predict possible required changes based on a patient's condition with a view to prevent severe health illnesses, facilitating decision support for caregivers [9, 10]. Furthermore, one of the most important capabilities that cloud presents is data storage; however, the acquisition of accurate diagnosis requires long-term monitoring, and hence the requirement of stronger cloud storage security mechanism to secure the collected data for long periods of time [11, 12].

2.1.4 Layer 4—Action

The two main objectives of this layer are to provide either 'active' or 'passive' actions. In the *active action*, an actuator is used to elaborate changes within the MCPS. For instance, a doctor sends an order to elevate the dose of medicine in the serum or the usage of robotic arms for *surgery assistance* [13]. Meanwhile, the *passive action* provides the opportunity of a better visualisation of the current state of the patient, allowing caregivers decision support.

2.2 Digital Twins

Digital Twin refers to the virtualisation of physical assets to monitor assets. This is enabled through electronic sensors communicating with each other using customised application software and are capable of controlling the digitised assets [14].

Nowadays, modern engineering technologies have immensely contributed to the evolution of healthcare services. In fact, the usage of mathematical models for processing of higher volumes of 'biodata' enables effective medical interventions. Some early prototypes of digital twins were realised in the ESB Logistics Learning Factory, which using a cloud-based software application a multidimensional data along with an information model were built [15]. By reading patients sensory data, even at the molecular level a 'digital' representation of a patient—a 'virtual patient' can be created. Therefore, the Digital Twin can be a platform to exhibit cutting-edge engineering solutions within health care. Many universities are training and preparing students in clinical technology, whilst doctors are working along with engineers from a wide range of backgrounds to enhance the functionality of contemporary medical practice [16].

In the wider context, Digital Twins are used to monitor the performance of artefacts and pieces of machinery in order to perform preventative maintenance. In fact, digitalisation of individual artefacts seems a straightforward task since it is based on the instrumentation of electronic sensors placed across the artefact. Besides, such artefacts have a unique shape after its manufacturing, making such instrumentation easier. In health care, however, the human structure is more complex due to the constant molecular and physiological changes throughout the body, making it very complex to extract precise molecular data. Unfortunately, Digital Twin seems a complex challenge, even though digital models of individuals' genetic, biochemical, physiological and behavioural features have already been implemented [17].

3 Impact of WBAN in Health Care

WBANs provide an extensive range of monitoring applications for different contexts such as health care, military, sports and video gaming, among others [18]. The healthcare sector is keen to adopt such a technology to enhance medical capabilities and to consequently improve the lifestyle of the human being. WBANs provide continuous health monitoring of patients allowing caregivers an easier decision support in order to send the necessary medical prescription or treatment without the requirement of patients' physical presence at clinics and hospitals [5]. WBANs used in health care consists of multiple *biomedical sensors*, which can be either wearable (e.g., fitness watches) or implanted under the human skin like electronic chips created with nanotechnology. Implanting chips under skin is referred to as '*biohacking*', a term that would be in discussion for the coming years with a focused consideration of the relevant cybersecurity challenges (e.g., cyber espionage). Some applications of WBANs are discussed in the following subsections.

3.1 Cardiovascular Application

The research from [19] shows that more than 20 million people suffer from cardiovascular diseases. The usage of WBANs allows to monitor users' health state remotely and on a real-time basis. Therefore, healthcare service providers can immediately prepare a preventive patient treatment plan when any abnormal information is measured by the sensor nodes.

3.2 Body Temperature

Body temperature is one of the most common physiological features measured through human activity monitoring [20]. It allows caregivers to detect medical stress that may lead to diverse health conditions based on the variation of corporal temperature. Such conditions include stroke, heart attacks and shock. Measuring body temperature is valuable to deter the physiological condition of a patient as well as for other care such as activity pattern monitoring [21, 22] and corporal heat harvesting [23].

3.3 Blood Glucose Monitoring

One of the current serious chronic health diseases throughout the world is diabetes. This disease has been increasing because of higher levels of sedentary habits given by comfortable options to the human being without having appropriate exercise. If diabetes is not treated properly and on time, it can cause serious complications such as blindness, stroke, kidney disease, heart disease and high blood pressure [24]. WBANs allow the continuous monitoring of patients' blood glucose level to provide information on healthier habits for food consumption and frequency of body exercise.

Currently, the measurement of blood glucose level can be done by means of a test strip pricking a finger. However, biomedical sensors can be implanted in a body to monitor the glucose level throughout the day. Using WBAN sensors, caregivers inject the necessary amount of insulin in patients using the actuator nodes within MCPS when the glucose level reaches a threshold [25].

3.4 *Stress Monitoring*

Stress leads to numerous diseases. It can lead to negative psychological illnesses such as anxiety, decreased patient satisfaction and depression [26]. Accelerated lifestyles in industrialised countries such as United States and Great Britain have increased stress levels among the population leading to negative consequences such as alcoholism and addictive smoking [27]. WBANs provide real-time monitoring of stress levels in individuals supporting physicians for appropriate treatments [28]. Modern smartphones can provide this service and the same platform can authenticate the users' fingerprints for privacy and security purposes.

3.5 *Rehabilitation and Therapy*

The main goal of rehabilitation is to support patients in restoring their physical and functional capabilities back to normal conditions, when they get dismissed from a hospital [29, 30]. Rehabilitation is a dynamic process in which necessary techniques are used to enhance the physical behaviour of a patient to his/her ideal physiological state. Therefore, tracking and detecting human mobility becomes an essential factor for home-based therapy treatments. Wearing or having implanted biomedical sensors, data fusion and real-time feedback for patients along with virtual reality environments are examples of techniques that could be used for rehabilitation [30].

4 Challenges in Medical Cyber-Physical Systems and Digital Twins

Even though these technologies are still in development and under specific research, it is suggested to study the challenges mentioned below in order to take the necessary actions to assure the appropriate balance between security, privacy and performance. The next subsections discuss some challenges on the studied technologies including cloud systems and Internet of things.

4.1 High Assurance Software

Software deployment is playing an increasingly important role when developing new MCPSs. Actually, the functionality of modern devices are software based and comparing with some years ago, when some functions were traditionally implemented in hardware have been replaced by software solutions. Thus, the higher demand for developing software offers confidentiality, integrity, reliability and ease of use to deploy safe and effective MCPSs and digital twins in the future. It is required to balance effectiveness in software engineering along with secure coding to avoid disruptions in healthcare organisations due to time consumption given by software patching.

4.2 Certification and Regulatory Issues

MCPSs and digital twins in health care are safety-critical systems and they must be prone to regulatory observation through certification or approval processes. Traditional regulatory regimes used by the Food and Drugs Association (FDA) to approve medical devices and medicines are becoming inappropriate due to the complexity of these cutting-edge technologies [31]. The FDA currently requires study cases as elements of the documentation submitted for future approval assurance considering regulatory modifications, for instance, the infusion pump improvement initiative [32]. Therefore, it is expected that similar or even specific and complex requirements will be demanded for the approval of MCPSs and digital twins, in general.

In addition, other important part of this challenge is software certification and methods to make it part within the regulatory approval process for the deployed device. Most of the medical devices possess large amounts of embedded software performing various monitoring and care delivery tasks. Considering that medical devices are becoming more complex and interconnected, it should become more evident for the requirement of certification and regulation at early design stages. This can be done in two ways: (1) The 'design for verification' approach [33] can support on better verification techniques including scalability and easier verification evidence generation; (2) model-based generative techniques can be used to perform verification early in the design and then extend the guarantees provided by the performed verification prior to its implementation through code verification.

4.3 Security and Privacy

These technologies provide interoperability capabilities, allowing to connect and transfer information through multiple platforms, acquiring functionalities that previously were never possible to appreciate; however, they also open new concerns in terms of security and privacy [34]. An attacker able to penetrate MCPSs or digital twin software has the potential and capability to harm or terminate the life of the patient by reprogramming devices [35]. There are four types of targets when attacking these systems [36].

- *Patient*: The attacker can target directly to the patient's health. It means attacking to the sensing, processing communication and treatment delivery aspects of the MCPS. For instance, reprogramming an infusion pump to provide a larger amount of medicine than the prescribed.
- *Data*: An attacker can access highly confidential and sensitive data belonging to the patient or the involved ones in the medical treatment. The loss of data privacy can lead to potential blackmailing, computer abuse and discrimination [37].
- *Device*: An attacker can perform a Denial of Service (DoS) attack on the MCPS, or also be part of it (e.g., wearable or implanted device), and deploy it to belong to a huge botnet in order to perform robust Distributed Denial of Service attacks (DDoS). Moreover, this can also result in privacy loss over systems that should be designed to fail open as suggested [38].
- *Institution*: The goal of a cyberattack is to compromise the interaction between the MCPS and the corporate network of the institution in order to obtain unauthorised access or at larger scale, patient data theft or network operational information infiltration.

Recent years have been a great issue for medical devices in terms of security addressed to several devices such as wearable, implantable [36, 38] or interoperable devices [37]. Nonetheless, in most of the cases, the focus is addressed to specific features of MCPS security like encrypted communication and effective access controls. In addition, the main challenge of deploying secure MCPS involves flexible and open solutions while mitigating the following issues: (1) heterogeneity of systems, (2) improving usability (even transparency) of security solutions developed and (3) considering safety implications of security solutions and decisions including the mitigation of human error and insider threats.

4.4 Challenges in Involved Systems

As studied in previous sections, thanks to the usage of WBAN, healthcare systems show dependency during the communication between patients and caregivers through cloud and IoT-based systems. The following subsections will discuss some challenges on the mentioned ones.

4.4.1 Cloud Computing

Cloud services are commonly available to users through the Internet (e.g., web browser) [39], using standard protocols and mechanisms for its communication [40].

External cloud communications are similar to any other communications over the Internet (i.e., traditional data centres). Therefore, the challenges faced by the cloud are the same as conventional IT solutions [41], including denial of service, Man-In-The-Middle (MITM), eavesdropping, IP-spoofing and masquerading attacks [42, 43]. Traditionally, these challenges are solved as the common ones such as implementation of Secure Socket Layer (SSL), IPSec, cryptographic algorithms, intrusion detection and prevention systems and digital certificates [42, 44].

Users and system administrators must be aware that cloud computing systems result in the sharing of computational, storage and network infrastructure resources [45], leading it to third-party risks. Shared network components allow attackers the possibility to perform horizontal privilege escalation techniques and the exploitation of other systems prior to the main target [46]. Commonly, users on cloud environments are granted with superuser privileges for the main purpose of managing their Virtual Machines (VMs) [47], and therefore attackers are motivated to acquire essential components from the system like IP and MAC addresses and perform malicious actions such as sniffing and spoofing over the real network.

The two main components of cloud are virtualisation and storage. Virtualisation allows the sharing of the same physical resources with multiple system environments. A separate VM is isolated for each user providing a virtual operating system, and the module in charge of managing the VMs along with the assigned resources is the VM Monitor (VMM) or hypervisor, allowing to run multiple operating systems at the same time [48]. Security challenges in terms of virtualization involve VM image sharing [46], VM isolation [40, 49], VM migration [46] and hypervisor issues where a compromised one can put all the VMs under the attacker's control [50]. Cloud system providers do not deliver to users full control over data, and users experience some control levels only on the VMs [51]. The fact that users do not have control over data belonging to the organisation results on significant third-party risks like data breaches. Moreover, the storage present in cloud environments shows challenges in terms of data privacy and integrity because data present in cloud is more prone to risks attempting against the confidentiality, integrity and availability compared to traditional data centre architectures [52]. In addition, data backup is an important element when having cloud systems and it demands to be secured against unauthorised access and illegal manipulation [40]. Generally, the access to cloud systems are done via web applications such as a Google Chrome for instance, and therefore, the requirement to protect these from vulnerabilities published by the OWASP [53].

4.4.2 Internet of Things

IoT is growing steadily and the medical sector is expected to experience an expanded adoption of it, creating cutting-edge eHealth IoT devices along with embedded applications. The challenges that IoT has for a secure healthcare networked environment include the computational limitations that devices present with their low-speed processors, memory and energy limitations. IoT networks present challenges in terms of scalability because of its high acquisition, along with the required compatibility with known network protocols. Medical devices are connected through several wireless protocols such as Zigbee, WiFi, GSM, WiMax, 6LowPAN, 3G/4G and soon 5G networks. The requirement of having a cross-platform system allowing IoT devices communicate with IP networks and making it part of known systems is a challenge as well, and another important aspect is the capability of producing tamper-resistant packets [54]. In-transit and stored health information can be eavesdropped or manipulated by an attacker. Some attacks include DoS attacks causing interruption, data breaches affecting the patient's privacy, data tampering and modifying the behaviour of sensing and delivering devices [55, 56].

5 Conclusion and Further Research

In this paper, the MCPSs and digital twins were studied and analysed as new incoming technologies that enhance steadily the capabilities of healthcare services. Medicine is integrating the usage of information technology (i.e., EHRs) and is keen to involve operational technology as well to raise the possibilities for a better lifestyle to patients. Even though the digital twin is not yet implemented in medicine, it is a topic that is worth to undertake research addressed to this field. WBAN is allowing to develop sophisticated MCPSs and would be of great support to deploy digital twin solutions as well; however, it also requires a deeper research in terms of the mentioned security and privacy challenges considering the integration of WBAN with IoT networks and cloud environments. Healthcare organisations and providers are keen to enhance their security maturity, hence the need for researchers to focus on the different systems and architectures in order to develop appropriate measures for this cutting-edge technology. In fact, it is recommended to critically analyse the challenges of big data platforms as well, and the possibility to deploy customised systems addressed to the vulnerability assessment of MCPSs in order to deploy the necessary security updates and bug fixes. Health care must be part of the CNI for all countries due to the level of extortion that a cyberattack can cause, leading to life or death decisions that caregivers could make during a system disruption. Therefore, the requirement to develop the next generation of security researchers to enhance the posture and the assurance of system and patient data.

References

1. Lee, I., Sokolsky, O., Chen, S., Hatcliff, J., Jee, E., Kim, B., King, A., Mullen-Fortino, M., Park, S., Roederer, A., Venkatasubramanian, K.K.: Challenges and research directions in medical cyber–physical systems. Proc. IEEE **100**(1), 75–90 (2012)
2. Movassaghi, S., Abolhasan, M., Lipman, J., Smith, D., Jamalipour, A.: Wireless body area networks: a survey. IEEE Commun. Surv. Tutor. **16**(3), 1658–1686 (2014)

3. Kocabas, O., Soyata, T., Aktas, M.K.: Emerging security mechanisms for medical cyber physical systems. IEEE/ACM Trans. Comput. Biol. Bioinform. **13**(3), 401–416 (2016)
4. Cavallari, R., Martelli, F., Rosini, R., Buratti, C., Verdone, R.: A survey on wireless body area networks: technologies and design challenges. IEEE Commun. Surv. Tutor. **16**(3), 1635–1657 (2014)
5. Anwar, M., Abdullah, A.H., Qureshi, K.N., Majid, A.H.: Wireless body area networks for healthcare applications: an overview. Telkomnika **15**(3), 1088–1095 (2017)
6. Babu, S., Chandini, M., Lavanya, P., Ganapathy, K., Vaidehi, V.: Cloud-enabled remote health monitoring system. In: 2013 International Conference on Recent Trends in Information Technology (ICRTIT), pp. 702–707. IEEE (2013)
7. Soyata, T., Muraleedharan, R., Funai, C., Kwon, M., Heinzelman, W.: Cloud-vision: real-time face recognition using a mobile-cloudlet-cloud acceleration architecture. In: 2012 IEEE Symposium on Computers and Communications (ISCC), pp. 000059–000066. IEEE (2012)
8. Powers, N., Alling, A., Osolinsky, K., Soyata, T., Zhu, M., Wang, H., Ba, H., Heinzelman, W., Shi, J., Kwon, M.: The cloudlet accelerator: bringing mobile-cloud face recognition into real-time. In: 2015 IEEE Globecom Workshops (GC Wkshps), pp. 1–7. IEEE (2015)
9. Mao, Y., Chen, Y., Hackmann, G., Chen, M., Lu, C., Kollef, M., Bailey, T.C.: Medical data mining for early deterioration warning in general hospital wards. In: 2011 IEEE 11th International Conference on Data Mining Workshops (ICDMW), pp. 1042–1049. IEEE (2011)
10. Kocabaş, Ö., Soyata, T.: Medical data analytics in the cloud using homomorphic encryption. In: Handbook of Research on Cloud Infrastructures for Big Data Analytics, pp. 471–488. IGI Global (2014)
11. Nalinipriya, G., Kumar, R.A.: Extensive medical data storage with prominent symmetric algorithms on cloud-a protected framework. In: 2013 IEEE International Conference on Smart Structures and Systems (ICSSS), pp. 171–177. IEEE (2013)
12. Hani, A.F.M., Paputungan, I.V., Hassan, M.F., Asirvadam, V.S., Daharus, M.: Development of private cloud storage for medical image research data. In: 2014 International Conference on Computer and Information Sciences (ICCOINS), pp. 1–6. IEEE (2014)
13. Barbash, G.I., Glied, S.A.: New technology and health care costs—the case of robot-assisted surgery. N. Engl. J. Med. **363**(8), 701–704 (2010)
14. Brenner, B., Hummel, V.: Digital twin as enabler for an innovative digital shopfloor management system in the ESB logistics learning factory at Reutlingen-university. Procedia Manuf. **9**, 198–205 (2017)
15. Bruynseels, K., Santoni de Sio, F., van den Hoven, J.: Digital twins in health care: ethical implications of an emerging engineering paradigm. Front. Genet. **9**, 31 (2018)
16. Uhlemann, T.H.J., Schock, C., Lehmann, C., Freiberger, S., Steinhilper, R.: The digital twin: demonstrating the potential of real time data acquisition in production systems. Procedia Manuf. **9**, 113–120 (2017)
17. Uhlemann, T.H.J., Lehmann, C., Steinhilper, R.: The digital twin: realizing the cyber-physical production system for industry 4.0. Procedia CIRP **61**, 335–340 (2017)
18. Qureshi, K.N., Abdullah, A.H., Anwar, R.W.: The evolution in health care with information and communication technologies. In: Proceeding of 2nd International Conference of Applied Information and Communications Technology-2014. Elsevier, Oman (2014)
19. Ullah, S., Khan, P., Ullah, N., Saleem, S., Higgins, H., Kwak, K.S.: A review of wireless body area networks for medical applications (2010). arXiv:1001.0831
20. Mukhopadhyay, S.C.: Wearable sensors for human activity monitoring: a review. IEEE Sens. J. **15**(3), 1321–1330 (2015)
21. Parkka, J., Ermes, M., Korpipaa, P., Mantyjarvi, J., Peltola, J., Korhonen, I.: Activity classification using realistic data from wearable sensors. IEEE Trans. Inf Technol. Biomed. **10**(1), 119–128 (2006)
22. Winkley, J., Jiang, P., Jiang, W.: Verity: an ambient assisted living platform. IEEE Trans. Consum. Electron. **58**(2) (2012)
23. Leonov, V.: Thermoelectric energy harvesting of human body heat for wearable sensors. IEEE Sens. J. **13**(6), 1–8 (2013)

24. W H Organization. Global report on diabetes (2016)
25. Schwiebert, L., Gupta, S.K., Weinmann, J.: Research challenges in wireless networks of biomedical sensors. In: Proceedings of the 7th Annual International Conference on Mobile Computing and Networking, pp. 151–165. ACM (2001)
26. Regehr, C., Glancy, D., Pitts, A., LeBlanc, V.R.: Interventions to reduce the consequences of stress in physicians: a review and meta-analysis. J. Nerv. Ment. Dis. **202**(5), 353–359 (2014)
27. Cassel, J.: Physical illness in response to stress. In: Social Stress, pp. 189–209. Routledge (2017)
28. Milenković, A., Otto, C., Jovanov, E.: Wireless sensor networks for personal health monitoring: issues and an implementation. Comput. Commun. **29**(13–14), 2521–2533 (2006)
29. Hadjidj, A., Souil, M., Bouabdallah, A., Challal, Y., Owen, H.: Wireless sensor networks for rehabilitation applications: challenges and opportunities. J. Netw. Comput. Appl. **36**(1), 1–15 (2013)
30. Zhou, H., Hu, H.: Human motion tracking for rehabilitation—A survey. Biomed. Signal Process. Control **3**(1), 1–18 (2008)
31. High Confidence Software and Systems Coordinating Group, B High-confidence medical devices: Cyber-physical systems for 21st century health care. A research and development needs report, NCO/NITRD (2009)
32. Goodman, C.: Food and Drug Administration Center for Devices and Radiological Health (1988)
33. Alexander, K., Clarkson, P.J.: Good design practice for medical devices and equipment, Part II: design for validation. J. Med. Eng. Technol. **24**(2), 53–62 (2000)
34. Ackerman, M.J., Filart, R., Burgess, L.P., Lee, I., Poropatich, R.K.: Developing next-generation telehealth tools and technologies: patients, systems, and data perspectives. Telemed. e-Health **16**(1), 93–95 (2010)
35. Halperin, D., Heydt-Benjamin, T.S., Fu, K., Kohno, T., Maisel, W.H.: Security and privacy for implantable medical devices. IEEE Pervasive Comput. **1**, 30–39 (2008)
36. Arney, D., Venkatasubramanian, K.K., Sokolsky, O., Lee, I.: Biomedical devices and systems security. In: 2011 Annual International Conference of the Engineering in Medicine and Biology Society, EMBC, pp. 2376–2379. IEEE (2011)
37. Venkatasubramanian, K.K., Gupta, S.K.S., Jetley, R.P., Jones, P.L.: Interoperable medical devices. IEEE Pulse **1**(2), 16–27 (2010)
38. Denning, T., Fu, K., Kohno, T.: Absence makes the heart grow fonder: new directions for implantable medical device security. In: HotSec (2008)
39. Kifayat, K., Merabti, M., Younis, Y.A.: Secure Cloud Computing for Critical Infrastructure: A Survey (2012)
40. Subashini, S., Kavitha, V.: A survey on security issues in service delivery models of cloud computing. J. Netw. Comput. Appl. **34**(1), 1–11 (2011)
41. Ficco, M., Rak, M.: Stealthy denial of service strategy in cloud computing. IEEE Trans. Cloud Comput. **3**(1), 80–94 (2015)
42. Sankar, K., Kannan, S., Jennifer, P.: On-demand security architecture for cloud computing. Middle-East J. Sci. Res. **20**(2), 241–246 (2014)
43. Liu, B., Bi, J., Vasilakos, A.V.: Toward incentivizing anti-spoofing deployment. IEEE Trans. Inf. Forensics Secur. **9**(3), 436–450 (2014)
44. Ali, M., Khan, S.U., Vasilakos, A.V.: Security in cloud computing: opportunities and challenges. Inf. Sci. **305**, 357–383 (2015)
45. Fernandes, D.A., Soares, L.F., Gomes, J.V., Freire, M.M., Inácio, P.R.: Security issues in cloud environments: a survey. Int. J. Inf. Secur. **13**(2), 113–170 (2014)
46. Hashizume, K., Rosado, D.G., Fernndez-Medina, E., Fernandez, E.B.: An analysis of security issues for cloud computing. J. Internet Serv. Appl. **4**(1), 1–13 (2013)
47. Bilal, K., Malik, S.U.R., Khan, S.U., Zomaya, A.Y.: Trends and challenges in cloud data centers. IEEE Cloud Comput. Mag. **1**(1), 10–20 (2014)
48. Neng-Hai, Y., Hao, Z., Xu, J., Zhang, W., Zhang, C.: Review of cloud computing security. Acta Electron. Sinica **41**(2), 371–381 (2013)

49. Gonzalez, N., Miers, C., Redigolo, F., Simplicio, M., Carvalho, T., Näslund, M., Pourzandi, M.: A quantitative analysis of current security concerns and solutions for cloud computing. J. Cloud Comput. Adv. Syst. Appl. **1**(1), 11 (2012)
50. Szefer, J., Keller, E., Lee, R.B., Rexford, J.: Eliminating the hypervisor attack surface for a more secure cloud. In: Proceedings of the 18th ACM Conference on Computer and Communications Security, pp. 401–412. ACM (2011)
51. Wei, L., Zhu, H., Cao, Z., Dong, X., Jia, W., Chen, Y., Vasilakos, A.V.: Security and privacy for storage and computation in cloud computing. Inf. Sci. **258**, 371–386 (2014)
52. Wang, C., Wang, Q., Ren, K., Cao, N., Lou, W.: Toward secure and dependable storage services in cloud computing. IEEE Trans. Serv. Comput. **5**(2), 220–232 (2012)
53. Owasp.org. https://www.owasp.org/images/7/72/OWASP_Top_10-2017_%28en%29.pdf (2018). Accessed 31 Dec 2018
54. Islam, S.R., Kwak, D., Kabir, M.H., Hossain, M., Kwak, K.S.: The internet of things for health care: a comprehensive survey. IEEE Access **3**, 678–708 (2015)
55. Zia, T., Zomaya, A.: Security issues in wireless sensor networks. In: Proceedings of the IEEE International Conference on Systems and Networks Communications, October 2006, p. 40 (2006)
56. Wang, Y., Attebury, G., Ramamurthy, B.: A survey of security issues in wireless sensor networks (2006)

Part II
Internet of Things, the Digital Twin Enabler

Present Scenarios of IoT Projects with Security Aspects Focused

Aniruddha Bhattacharjya, Xiaofeng Zhong, Jing Wang and Xing Li

Abstract To explore IoT's hidden prospective and to address many global complications, the International Telecommunication Union (ITU) is working. They are making the IoT standardized for several years in the Telecommunication Standardization Sector (ITU-T). ITU-T Study Group 20 was formed in recent times, to further endorse coordinated advancement of global IoT technologies, services, and applications. Some of the important IoT projects, their security pitfalls and their applications domains are discussed here. We need some secure architecture. Case by case, we need multi-layer architectures for secure IoT, such as in the smart city environs; we have numerous protocols, access technologies, functions and several types of nodes. Universally, future focuses in the security issues of the Internet of Things would typically quintessence on the following features, related laws for the security of the Internet of Things, the open security system, terminal security function, individual privacy protection mode, etc. We have developed a *Secure Hybrid RSA (SHRSA) messaging system* for End to End encrypted messaging, with solutions to many bottlenecks of RSA and Instant messaging schemes. Our scheme has much more decryption efficiency. Presently we have used our *Secure Hybrid RSA (SHRSA)* cipher for secure and efficient messaging scheme. We have found in real-time testing results analysis that, our scheme is much more authentic, efficient and secure system. So as a cipher *Secure Hybrid RSA (SHRSA)* cipher can be used in present IoT communications and in near future in Future Internet of everything (IoE) communications.

Jing Wang died before publication of this work was completed.

A. Bhattacharjya (✉) · X. Zhong · J. Wang · X. Li
Beijing National Research Center for Information Science and Technology, Department of Electronic Engineering, Tsinghua University, Beijing, China
e-mail: li-an15@mails.tsinghua.edu.cn

X. Zhong
e-mail: zhongxf@tsinghua.edu.cn

J. Wang
e-mail: wangj@tsinghua.edu.cn

X. Li
e-mail: xing@cernet.edu

M. Farsi et al. (eds.), *Digital Twin Technologies and Smart Cities*,
Internet of Things, https://doi.org/10.1007/978-3-030-18732-3_7

Keywords IoT@work · IoT-A · BeTaaS · OpenIoT · IoT@work · Secure hybrid RSA (SHRSA) · SHRSA encryption · SHRSA decryption

1 Introduction

The Internet of Things (IoT) [1–20] signifies the interconnection of exceedingly heterogeneous networked entities for instance sensors, actuators, smart phones, etc. At first, let's highlight on the ongoing projects [8, 11, 21–30] and consider them as our case studies. In this chapter, we have discussed some of the important IoT projects like European Union FP7 project [22], Hydra [24], iCore [26], HACMS [27], National Science Foundation projects [28], and FIRE [29, 30] etc. Then we have highlighted their security features and have highlighted different domains of applications, then security pitfalls and last but not the least, have described our model followed by conclusions. We have highlighted our approach with its End to End security aspects.

In this chapter, Sect. 1's subsections have highlighted some of the significant important projects, which have some security pitfalls and some have good security (Sect. 1.1). Then another subsection of Sect. 1 has highlighted very important IoT projects around the world with some sort of structural analysis. The Sect. 2 has highlighted different practical implementations [31–46] domains of IoT projects nowadays. Section 3's subsections have highlighted two things-

1. Securities of existing IoT projects.
2. Privacy and trust 'techniques in IoT.

Our approach for providing End to End security in messaging scenario, which is a nine layer protocol stack, is described in Sect. 4. Our whole messaging scheme is named as *Secure Hybrid RSA (SHRSA) messaging scheme* [47–51]. This cipher-Secure Hybrid RSA (SHRSA) can be later used for data encryption in IoT and Internet of Everything (IoE) scenario. Here we have also discussed our scheme advantages than existing Instant Messaging schemes. Moreover, we have highlighted three major distinguishable aspects of our work, those are- *less memory occupancy, less CPU usage and much more efficiency in decryption process* [47–51]. Section 4 has concluded our chapter.

1.1 Present Ongoing Important IoT Projects

In this section, sub-sections have highlighted some of the significant important projects, which have some security pitfalls, and some have good security.

1.1.1 European Union Projects

The European Union is working on a project called as Butler (European Union FP7 project) [21]. This project facilitates the expansion of secure and smart life assistant applications, along with the security and privacy necessities. Also this work has developed a mobile framework. The smart applications which are targeted, are like smart-home/smart office, smart-mobility/smart transport, smart-health, smart-shopping, and smart-cities.

Another European Union project is EBBITS (EU FP7 project) [23]. This project works for an *Intrusion Detection System (IDS),* by use of latest IPv6 over 6LoWPAN devices. Ever since, 6LoWPAN protocol is defenceless to wireless and Internet protocol attacks [52]. This project has projected a IDS framework comprises of a monitoring system and a detection engine.

The Hydra project [24] has projected a middleware for Network Embedded Systems. This middleware is founded on a *Service-Oriented Architecture (SOA).* Hydra project has considered the distributed security concerns and social trust within the middleware constituent. Hydra [24] is designed for P2P communication and diagnostics, architecture is formed on Semantic Model and the Device and Service Discovery.

Another project which is to increase the user trust is uTRUSTit [23]. uTRUSTit [23] stands for, Usable Trust in the IoT (EU FP7 project). It is actually a trust feedback toolkit to potentially increase the user trust. It empowers the system manufacturers and system integrators, to express the security ideas. It has agreed to create effective decisions on the trustworthiness.

iCore [26] is another EU project. iCore [26] has a management framework with very significant security protocols/functionalities. These protocols/functionalities are having relation with the ownership and privacy of data and the access to objects. This management framework has three levels of functionality: virtual objects (VOs), composite virtual objects (CVOs), and functional blocks. The iCore solution can be part of various smart environs, like supply chain management, smart-office, smart-transportation, and ambient-assisted living.

1.1.2 DARPA and NSF Projects

Now very well-known Defense Advanced Research Projects Agency (DARPA) project is HACMS [27]. It stands for *High Assurance Cyber Military Systems.* This project actually has tried to have patch of the security vulnerabilities of IoT. This project takes account of drones, medical equipment, and military vehicles. HACMS [27] provides the seeds for future security protocols, achieves sufficient standardization and security.

National Science Foundation (NSF) has a *multi-institutional project* [28]. This project is actually working for the security in the cyber-physical systems. This multi-institutional project is working on several solutions, like trying to discover the efficient resolutions, finding novel network architectures and networking conceptions, trying to invent new communication protocols. They are bearing in mind about the trade-offs of between mobility and scalability, technical challenges, trusted data and the integrity. Along with that, they are also bearing in mind about authentication, trust models, and use of network resources on mobile environments.

1.1.3 EU, Chinese, Japanese and Korean Projects

The EU, China and Korea are working together in a project called FIRE [29, 30]. It stands for *Future Internet Research and Experimentation*. The FIRE [29, 30] works for discovering resolutions, for the setting out of IoT technologies in numerous application areas, like medical and health service, urban management, social security, people livelihood, public safety. They are also trying to give proper focus on intellectual property right, privacy and information security.

Another EU and Japan collaborative project is EU Japan ICT Cooperation project [22]. They have already made the common global standards, to make sure, about seamless communications and shared ways to accumulate and have right to use the information. They are also trying to confirm the of highest security and energy efficiency standards.

1.1.4 Digital Twin

Now we have more practical focused new technique called "*Digital Twin*". It is not an astonishing thing that most vendors of IoT Platforms have implemented some form of a digital twin. We can say in other word where digital twin models need data, IoT feeds those data. These are usually named as twins, shadows, device virtualization, etc. The term "Digital Twin" was defined by Dr. Michael Grieves at the University of Michigan around 2001–2002. He at the beginning defined this in the context of Product Lifecycle Management. Here we use three very specialized tools like-*conceptualization, comparison,* and *collaboration.* These three very special attributes contribute to make the foundation for the next generation of problem solving and innovation. The Digital Twin concept model is consisting of three main parts: *(a) physical products in Real Space, (b) virtual products in Virtual Space*, and *(c) the connections of data and information* that bind the virtual and real products together. But this twin model has a problem, when data are exchanged between Real space and Virtual space and information are processed, those data and information are totally insecure. So some sort of secure communications are needed here also. Any kind of *lightweight cipher* can be incorporated here to make a secure communication protocol for those data exchanges.

1.2 Security Features IoT Projects

Some of the well-known existing research projects are: Internet of Things at Work (IoT@Work) [11], Building the environment for the Things as a Service (BeTaaS) [23], Open source cloud solution for the Internet of Things (OpenIoT) and Internet of Things Architecture (IoT-A) [11]. These are some important case studies also [7, 11–19, 21–30]. Let's discuss one by one.

1.2.1 IoT-A

IoT-A (Internet-of-Things Architecture) [8, 11] is developed with an EU FP7 project until 2013. It is, an architecture reference model, advanced with already running community progress. This architecture actually uses the conceptions of views and perspectives to direct the generation of architecture cases, from business objectives via necessities. This kind of views and perspectives consist of the information view for static structures, along with dynamic information flows. Furthermore, it consists of the performance and scalability viewpoint, and the trust and security standpoint. Depend on the business objectives, the necessities are the outcomes, just from a multitude of unified necessities. Afterwards these necessities are transformed into fine-grained necessities, for an architecture instance. The unified necessities are presently 38 and they are focusing on the security and privacy viewpoints. Furthermore, IoT-A [8, 11] encompasses numerous models that are self-regulating from a specific architecture. These models have various types of models, like the communication model and the trust, security and privacy model etc.

IoT-A Reference Model as a Common Ground

Founding the common platform or ground, incorporates the explanation of IoT entities and description of the basic exchanges and relations with each other [31–46]. The *Architecture Reference Model (ARM)* is actually provided that same kind of common platform or ground for the IoT field. Hence, it is well understood that, any party envisioning to form an IoT system, which is *IoT-A (Internet of Things Architecture)* [8, 11] compatible, requisites to be built on the common conceptions, already present in the IoT-A Reference Model. IoT-A is the European Lighthouse Integrated Project and it has addressed for three years the Internet-of-Things Architecture.

One more advantage is that, we ca use the *IoT-A Architecture Reference Model (ARM)*, for the generation of compliant architectures for particular systems. This is very easy, just by use of tool support; we just have to enable this tool support.

The advantages of this kind of generation scheme for IoT architectures [1–30], are like it gives the *automatism* of this process, and as a result saving the R&D efforts. The created architecture will offer *intrinsic interoperability* of the resulting IoT systems. Another aspect is that, if we are using the above system-generation tools, totally modelled on the IoT-A ARM, then we have one outcome. Outcome fact is that, any variances in the derived architectures can be attributed to the discrimination of the relevant use case. In the case when we are applying the IoT-A ARM, then one thing is that, the estimates of system complexity, etc. are accessible, for the system parts to be implemented. So as some outcomes, the judgment process of the overall execution and implementation work for use case execution and implementation becomes much easier.

Moreover, another indirect good advantage is that, some of the projects that might not have been understood clearly, by reason of uncertainties in the project plan, can become possible to understand for implementing it. As a whole the total implementation effort is definitely *less* than developing an architecture, which is devoid of the help of an architectural reference model.

Another significant use can be *bench-marking*. One of the real-time example can be a reference architecture for new exploration vehicle, which was used by NASA. It was just to have better bench-marking tenders; it was going to obtain for the duration of a public bidding procedure. At the time the reference model recommends the language, for using in the systems/architectures to be evaluated, the reference architecture states the least (functional) prerequisite on the systems/architectures. By standardizing the explanation and also the arrangement and marking out of system components and facets, it also offers a high level of transparency and integral comparability to the bench-marking procedure.

Figure 1 has shown the high-level taxonomy of the discussed reference-architecture process. In the real-time scenario, the derivation of IoT-A-consenting domain-specific architectures from the reference architecture, is a best practice. Necessary inputs for defining the IoT reference model are, existing architectures, business scenarios, and stakeholder concerns. The most important and obligatory thing is that, to generate a common considerate of the IoT domain from the diverse inputs, it is nothing but a modelling exercise. In this process the specialists have to work in an organized way and they have to extract the key concepts and their associations of the IoT domain from existing knowledge. Moreover, stakeholder worries, in effect architectures, and business developments, can be converted into application-oriented necessities as shown in Fig. 1. When we try to get conclusions from those, then these necessities produce a set of unified necessities, or we can call it as *unified requirements* as shown in Fig. 1. *Unified requirements* then define the *IoT Reference Architecture* as shown in Fig. 1. In the interior of the *Architecture Reference Model (ARM)*, the IoT Reference Model controls the explanation of the IoT Reference Architecture [21–30]. It actually forms the dependencies among the *Reference Architecture* and *the Reference Model* as shown in Fig. 1. As soon as the change is projected in the Reference Model, then we can have a clear chain of dependencies and this can give a direction to have succeeding changes in the interior of the *Reference Architecture* as shown in Fig. 1. As an outcome, an overall consistency of the *IoT-A (Internet of Things Architecture) Architecture Reference Model (ARM)* is retained.

In our daily life, we have a requirement for a detailed architecture process that recognizes single tasks inside the development process. This gives the actual insight in the dependencies of the mentioned tasks, and that offers a *dynamic model* of the development process step by step. The *Architecture Reference Model (ARM)* development process comprises of one key process that is the *ARM derivation*. There

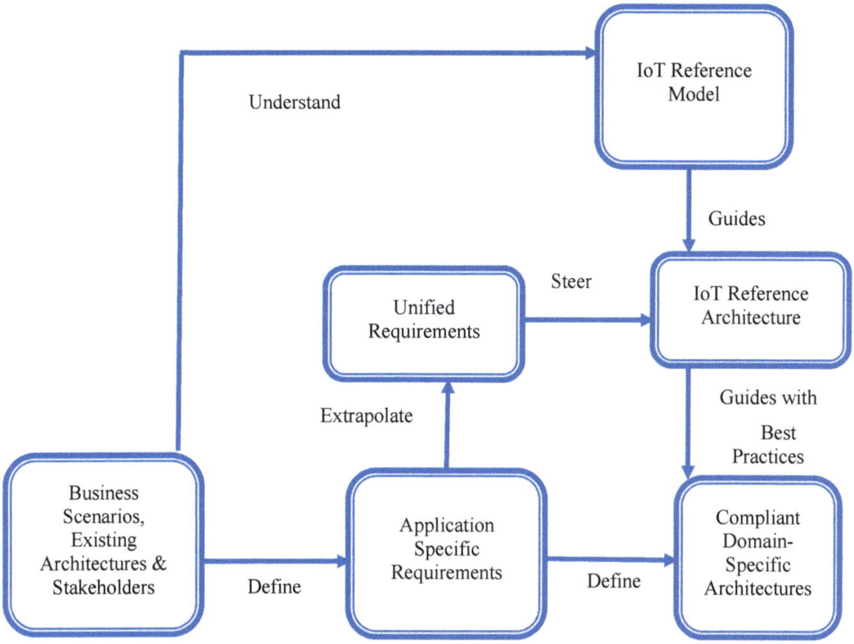

Fig. 1 High-level taxonomy of the IoT-Reference-Model and IoT Reference-Architecture dependencies and model effects

are two actions models, inside the ARM derivation, *the domain modelling* and *the functional modelling*. The *domain modelling* is responsible for forming the IoT Reference Model. The *functional modelling* is the key provider to the IoT Reference Architecture. This procedure accepts input from the requisite-collection procedure, which in sequence obtains input from external stakeholders and the state-of-the-art surveys executed for the period of the initial stages of IoT-A. In order to improve the impact of the *architectural reference model* in a best way, we have to recognize the circumstances, where IoT technologies have an exact importance. Here we already considered that, these scenarios regularly share the same users, stakeholders, sensors, and applications. In fact, the IoT Reference Model gives us the super vision, for the explanation of the IoT Reference Architecture *as shown in* Fig. 1.

Now let's consider some of the scenarios.

IoT-A [8, 11] encompasses five logical security constituents for addressing the security necessities. *Key Exchange and Management (KEM) component* is responsible for the Network security issue. KEM controls the cryptographic keys. These keys are actually being used for the authenticity, integrity and addition to those confidentiality also. For the resource constrained devices, IP Security (IPSec) tunnels among (unconstrained) gateways is used by the KEM. It's a very integrated conception to get the maximum coverage of network security. But one big issue is that, the connections among constrained devices and the gateway always been *defenceless*.

Moreover, KEM does not work on the obtain-ability in the perspective of network connections. Another good point is that the KEM also addresses functional necessities, for instance lawful interception.

IoT-A [8, 11] encompasses three modules those deal with the necessities of identity management. The module *Identity Management (IM)* actually focuses on the generic management, but does not focus on the specific security requisites. The *module Authentication (AuthN)* deals with the authentication necessities for users and services. Also it deals with the accountability and non-repudiation. The *module Authorization (AuthZ)* deals with the authorization necessities for services, by use of the *role-based access control (RBAC)* along with *attribute-based access control (ABAC)*. Revocation is based on the specific access control model which is being used.

The privacy issue is managed by *Pseudonymisation (PN).* It uses the pseudonymization for services, users, and devices. Pseudonyms substitute real identities, which are obtained from KEM, but still retain pairing of identities and pseudonyms to guarantee accountability. Pseudonyms can additionally deliver unlinkability, given a new pseudonym for each and every action is used. But the Pseudonymisation (PN) does not deal with complete *anonymity* and *data privacy*. However, AuthZ offers some way to have the right to use the granularity that may resolve data privacy to a definite level.

The module *Trust & Reputation (TRA)* manages the trust obligation for entity and device trust. In specific, the module describes the gathering of the user reputation for doing the calculation of the service trust. *IoT-A (Internet of Things Architecture)* defines the fault handling model, or functional group correspondingly. Necessities and measures of this model comprises *repairing the system, spotting existing failures, decrease of effects of failures* and *forecasting possible failures*. Therefore, the first method deals with avoidance, while the latter three deal with a life-cycle for mitigation.

1.2.2 BeTaaS

IoT and Machine-to-machine (M2M) [14] communication can be done very well with one architecture called as *Building the environment for the Things as a Service (BeTaaS)* [7]. The architecture empowers running applications over a local cloud of gateways. The highlighting feature is that, each BeTaaS [7] instance forms its own *cloud of gateways* that incorporates numerous *heterogeneous M2M systems* in a seamless way. The *Things as a Service (TaaS) reference model* is the main formation inspiration for the BeTaaS [7]. The architecture encompasses of *four layers*. The *Physical Layer* which is the first layer, encompasses the M2M systems which are connected to the platform. The Second layer is the *Adaptation Layer*, it deals with the connection to the physical layer. Moreover, it works for abstracting from peculiarities of the each and every M2M systems [14]. The third layer the *TaaS Layer*, depends on the abstraction layer and offers network-wide right to use to the devices, which are the M2M layer. Last but not the least, the *Service layer* controls the functionalities and services of BeTaaS applications. But one highlighting issue is that, the BeTaaS architecture is dealing with the security necessities, by offering distinct mechanisms for all of its layers excluding the physical layer.

If we think about Network Security, the Key Management component work with that by associating entities, by executing authentication, managing user sessions, and offers encrypted communication. Meanwhile the BeTaaS [7] instances comprise of various gateways. The BeTaaS makes use of the public key infrastructure (PKI) along with a Certificate Authority (CA) to accomplish keys and guarantee integrity, authenticity and confidentiality through the secure communication channels. BeTaaS also can work with circumstances, where several involved organizations, e.g., external entities that are not administered by the internal CA. This kind of cross-organization key management is managed by the BeTaaS directory service. Moreover, BeTaaS deals with resourced constrained devices by making use of the computationally more efficient cryptographic schemes for instance *Elliptic Curve Cryptography (ECC).*

For Identity Management, BeTaaS [7] offers authentication by making use of a *dedicated architectural component*. For this purpose, the dedicated architectural component separates two circumstances: *gateway level authentication* and *service level authentication.* In case of the gateway level authentication, the gateway joins a BeTaaS instance, and in case of application or service level authentication, a user uses an application. In case of the first circumstances, the authentication module makes use of the key management, while for the latter case; *OAu*th can be taken into consideration for authentication and authorization. Authorization is protected by a dedicated component as well. But one disadvantage is that the accountability obligation still unclear.

As we know that, the Privacy is specified as an important feature of the security procedures in BeTaaS, but there is no indication of how this prerequisite is achieved. Managing the distinctiveness of sensors and gateways, are the main work responsibilities of the identity management component. But no care about data anonymity or pseudonymity is here. *The trust and reputation component* is responsible for the dealing of the Trust issue. The model actually works like it, tries to get input from distinct trust characteristics: security mechanisms (for example it comprises of information concerning the encryption algorithms, the certificates, etc.), Quality of service (QoS) satisfaction, dependability enactment, battery load and stability in supplied data. These trust features are then collected to calculate the final trust value.

Here the resilience is managed with four dissimilar pillars: *fault prevention, removal, tolerance* and *forecasting.* Finding the reason for the potential reasons of failures and for offering resolutions to appropriately controlling them, is the works of the *Failure Analysis Approach.* The Failure Modes Effects and Critically Analysis process is executed on the functional items of the system. Here at the beginning, the fault modes for each IoT device are identified and corresponding effort on the analysis and operations is computed. Moreover, after evaluating the probability of failure happening, it allocates the criticality of the failure. Now here, the Reliability Architectural Approach module intends to offer resolutions for resolving the likely system failure, relating to the above-mentioned analysis.

1.2.3 OpenIoT

The EU FP7 OpenIoT research project (2012–2014) [23], has projected an IoT architecture which is formed on IoT-A's well-defined *Architectural Reference Model (ARM)*. It takes the key ARM ideas and functional building blocks. Nevertheless, OpenIoT focuses on offering a cloud-based middleware infrastructure. Therefore, this architecture can offer an on-request right to use the IoT or the IoT services, which is actually framed over several infrastructure suppliers. OpenIoT also proposals an open source implementation that is mainly focused on forming principles for the IoT applications, with the use of cloud-based characteristics for instance *on demand* or *pay-as-you-go service delivery*. So in a nutshell, the architecture pacts with IoT/cloud convergence. The OpenIoT architecture description defines two security modules, the *security & privacy module* and the *trustworthiness module*. Inside the security module, one sub module works with secure messaging and the other one works with authentication and authorization. But one thing is to mention is that, the privacy features are not in existent in the public code, which is not as per the specification. The trustworthiness module works for evaluating the trustworthiness of sensor data, which are taken as input (data trust).

OpenIoT [23] depends on the HTTP along with the TLS protocol, for ensuring secure and encrypted messaging. OpenIoT uses a *centralized security* and *privacy module* for identity management, which offers authentication and authorization with the help of OAuth. Here for managing the authorization, the role-based access control (RBAC) model is being used. One point to make clear is that, the trust module is an independent module in OpenIoT. The trust module works for the provisions of trust for both, data and device. To attain the device trust, spatial correlation of sensors is used by the OpenIoT. One of the examples can be, close sensors in alike environs always must yield alike sensor readings. At the time when the device trust is established, data records can be marked up with the trust labels.

1.2.4 IoT@Work

For establishing an IoT architecture for the industrial automation domain, one of the very renowned project is IoT@Work [23]. It is a European Commission FP 7 project, which was completed in 2013. The project was started to deal with security, auto-configuration, and interoperable and reliable network communication. For resolutions to deal with those, IoT@Work [23] brings together several concepts. Some of the example can be the concept of network slices, a combination of virtualization, resource management, and security. In real time, IoT@Work is handling network security via usually used technologies. *Extensible Authentication Protocol (EAP)* as an IEEE 802.1X implementation, guarantees authentication in the low network layer, like for switch ports. EAP-TLS also safeguards the confidentiality feature. The concept of network slices is allowed for being virtual in the network, and as a result, fast network link fail-over to defend availability. At the same time IoT@Work takes care about the device integrity but not about network integrity mechanisms.

Authentication is primarily offered by network security in IoT@Work. Additionally, authorization is achieved by *Capability-Based Access Control (CBAC).* Moreover, it supports for revocation, accountability, and delegation. CBAC has some good advantages like it works perfectly with many entities along with the situation when connection failure occurs to the central authorization service.

IoT@Work does not give more focus on privacy for the reason; it actually has main focus on the *industrial automation.* Nevertheless, some data privacy is offered with the modelling of granularities in access abilities. Moreover, for pseudonymous right to use by delegating abilities to a pseudonym, the access delegation tactic can be a very good option for use. Henceforth, entities can manage the anticipated level of unlink-ability on their own. Nevertheless, no clear provision for unlink-ability is specified. Using the *Zero Knowledge Proofs (ZKPs)*, the ability for getting the anonymity can be attained. As a result, no identifiers must be displayed for getting the right to use a device or service. One point to make clear is that, IoT@Work does not have mechanism to defend the trust-based requirements. Resilience with main focus on failure handling is a core prerequisite for IoT@Work. The virtual network links are used by the network slice method, that are forceful against failures. Furthermore, live-reconfiguration is conceivable and therefore permits for recovery in the sense of resilience. Though, the IoT@Work keeps a strong network emphasis and almost not focus on the devices and services.

The IoT@Work architecture has been advanced through an agile process, as depicted in the Fig. 2, which can be considered as the *model-driven architecture development method.*

The initial point for the architecture design is nothing but with making use of *scenario-driven requirements*. These scenarios used to form the *system model.* This system model presents that how the Internet of Things is anticipated to affect the factory and automation systems precisely in a generic way. These necessities are also nothing but displays of the specificities and constraints of current systems. Here along with that *top-down architecture design methodology*, an early technology testing activity has also been started for the purpose of the deeper considerations of the available methods and techniques and the higher-level abstractions, these can upkeep. The *technology testing* activity is a bottom-up design methodology. This methodology permits testing the present technologies with regards of satisfaction of IoT@Work architecture [23] necessities or of outlining the desirable extensions.

Currently the project has offered a new way to form an *IoT reference model,* typically with the specifications of the IEEE Standard 1471–2000 software architecture approvals. The IoT-A project just make an use of the diversified visions of the all stakeholders of an IoT-system. So as a result, we actually get quite a few models comprising an IoT functional model. Depend on the classical ISO-OSI layer model, the latter model collects the functions to form the functional groups deprived of, where the function groups are not essentially layered. Therefore, this kind of alike method is implemented in the IoT@Work project. Here actually, grouping is done as per the functionalities for particular functions consistent with what their specialism. A practical example can be like the *group of functions,* who are responsible for making the network of things running and works for adjusting the communication as per the

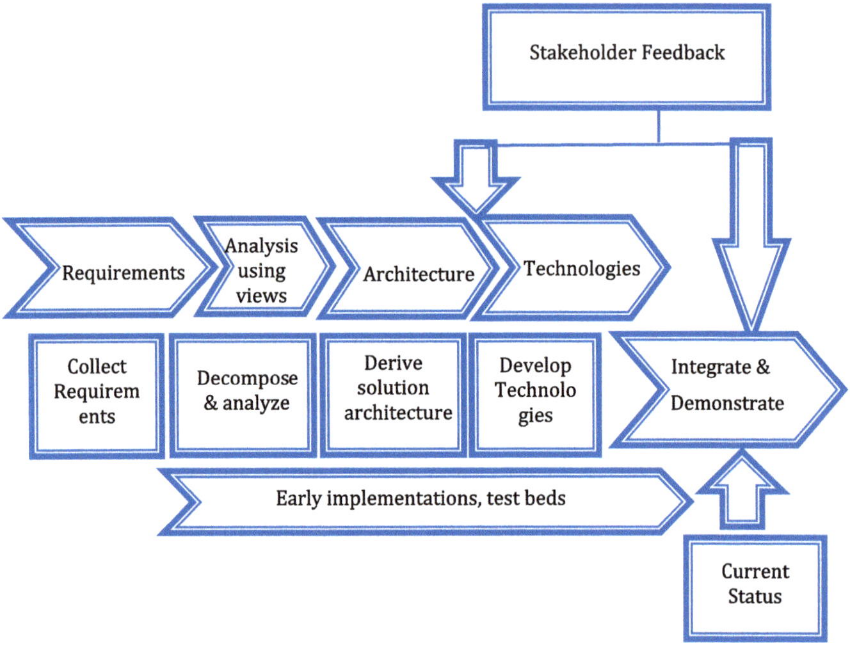

Fig. 2 IoT@Work architecture definition process

application requirements. Another function group works for *handling application layer actions* created by things and it's as per the logic of the application.

The architecture which is the outcomes from this is comprises of numerous functions, which are applied by numerous constituents. Here the cross-cutting issues are organized into planes and they are orthogonal to the layers. The layers are well-defined as *abstractions* and *function groups*. Hence, it is well understood that, these layers are responsible for overall management for handling the IoT infrastructure from the lowermost layer to the top most layer where, IoT applications run. Furthermore, another focusing point is that, among these two, the function groups comprise *management* and *orchestration functions*. These functions work for the formation and also works for application's constant running, on uppermost part of the resources and services existing in the IoT infrastructure. The functional grouping projected with three functional layers as follows:

1. The *device and network embedded services*. It is the first abstraction of the infrastructure, along with associated management functions also. These functions comprise securing physical constituents, managing communication interfaces, allocating identifiers and accumulating device semantics and context etc.
2. The *second abstraction layer* works for handling embedded resources and services with a special type of policy like in a sum up way. Moreover, it works for hiding some of the specifics of single constituents or devices. Here the func-

tions comprise security administration, network abstractions, low-level system observing and service directories.

3. The *third layer of abstraction* provisions straightway the application with the use of particular middleware facilities, which are for IoT setups. As this architecture is exclusively for the automation field, so these functions comprise a messaging bus, application resource explanations (e.g., ask for trustworthy communication or security setting is interpreted here). The *application logic* is interpreted at formation or runtime. Also, the interfaces to the dissimilar IoT management constituents are well-defined here. *Semantic reasoning functions*, along with other supportive functions can be also put here.

The IoT-centered architecture is well-defined inside the area of automation systems. Hence, there is a concentration on those functional parts that should offer trustworthy and communication with security, which is obligatory by some automation applications. The IoT tactic to the embedded systems is depend on the model, that virtual and physical are interlinked and reinforced by self-managing features of the Internet protocols. Numerous functions and resources obtainable by embedded devices (subset of smart objects or things), can be encapsulated into virtual objects. These are invoked or made accessible to a range of applications and services, which contended to have the right to use and use the things, for example, their physical and virtual resources. An IoT architecture in point of fact has to afford *trustworthy communication* and *assured security*, as per the automation systems requisite it.

The main focusing areas of the IoT@Work methodology are:

1. Assured Quality of service (QoS) using resource reservations as opposed to relative priorities.
2. Decoupling of concept and implementation is another focused work. Here except DiffServ/IntServ, IoT@Work do not blend the QoS abilities of the IP or Ethernet layer with the interface to the higher layers. Therefore, they can afford a solution, which works for a wide variety of technologies and topologies, along with present Industrial-Ethernet standards. It also means that, this architecture can work over a mix of layer-3 and layer-2 networks.
3. Central management considers real resource obtainability along with application networking requirements for instance QoS, trustworthiness and service reputation. Central management can offer optimizations that are very hard to accomplish, by use of a distributed hop-by-hop reservation system. We should know that, central management does not essentially necessitate a central implementation, distribution using domain controllers and device agents. It is promising and can be well-thought-out. A slice can be well-defined for collections instead of just-per-flow reservations.
4. Another main focusing area is Path manipulations; it is another way for accomplishing the traffic engineering objectives. Here the customer of the network takes his/her own path selection judgments.

2 Other IoT Application Areas

2.1 Transportation/Logistics

In transport logistics, IoT progresses the material flow systems. Moreover, it advances the global positioning and auto identification of freights [2]. Furthermore, it upsurges the energy efficiency and as an outcome, it cuts the energy consumption. In a nutshell, in intelligent cargo movement, by use of IoT, can make revolution, in the global supply chain. This revolution can be accomplished by dint of nonstop process synchronization of supply-chain information, and continuous real-time tracking and locating of objects. As an outcome, the supply chain will be controllable nature, noticeable and transparent and it will empower intelligent communication among people and cargo.

2.2 Smart Home

In near future the smart homes will consider mainly three issues: the *real time resource usage scenario* (for example water conservation and energy consumption), *security issues*, and *comfort issues*. Our smart home objective should be attaining better levels of comfort, even though cutting the overall expenditure. At the same-time, the smart homes also should deal with the *security* issues. It should have *complex security systems* for identifying the theft, fire or illegal entries in the inside of smart home. The participants included in this scenario form a very heterogeneous group [19, 53]. It is very well understood that, in this kind of scenario, there are dissimilar players that will collaborate in the user's home, for instance media-service suppliers, electric-service companies, telecommunications operators, Internet establishments, device makers, security firms etc.

2.3 Smart City

In a generic way, we understand that, the smart city highlighting area [13] will be living, governance, environment, economy, mobility and off-course people. Strong human along with social capital and ICT set-up a boost for all these smart city highlights. If we consider an example of a city of 1 million people, in next stages, a first business scrutiny determines that, numerous sectors/industries will get direct advantage from more and more digitalized and intelligent cities like as follows-

- Smart metering, 600.000 m, US $120 million prospect.
- Infrastructure for recharging electric vehicles, 45.000 electric vehicles, US $225 million prospect.

- Remote patient monitoring (diabetes), 70.000 people, US $14 million prospect.
- Smart retail, 4.000 stores, US $200 million prospect.
- Smart-bank branches, 3.200 PTMs, US $160 million prospect.

2.4 Retail

The customer requirements and business prerequisites are both the matters, IoT has to realize. One of the examples can be Price evaluation and identifying the differences of a product. Another example can be, finding for further goods of the same class but much cheaper. Therefore, it's well understood that, having this information in real time, benefits enterprises for advancing their business and to fulfil the customer requisites. It's well understood that, the existing big retail chains will try their best to take benefit of their leading position with the intention of enforcing the future IoT retail market. The same thing happened in that past like, as it occurred with RFID acceptance, which was applied by WalMart in 2004. Mainly, companies with governing spots, for instance Metro AG, WalMart, Carrefour, etc. are capable to push the acceptance of IoT technology due to their considerable market power.

2.5 E-Health

The main objective of the e-Health in near future will be controlling and stopping health problems. Nowadays we already have wearable tracking devices, so it's well understood that, we can have the option of being tracked and monitored by consultants even though locations are not same [3]. Health history of the marked peoples is another point that transforms the IoT-aided eHealth very versatile. So in near future we can have lots of business applications, that could give proposition for the opportunity of medical service for the patients as well as for the specialists, who requisite information to carry on their medical assessment. So it's well understood that, IoT makes human interaction much more capable with very high efficiency, as it not only empower localization, but also empower the tracking and monitoring of patients. An example can be, supplying information about the status of a patient, automatically result in more efficient process and makes people much more contented. The most significant stakeholders in this situation will be public, private hospitals and institutes along with that its well-established fact that the telecommunications operators are moderately active in e-health. But system security and *operational information security* are still a big problem [3, 4].

2.6 Environment

In the environmental domain, applications have numerous commonalities with other set-ups. Example can be *smart home* and *smart city*. The significant matter in these circumstances is to identify the way to save more and more energy. *Smart Grid* is the one of the most projecting domain in recent time. We need to take lots of initiatives that will entail a much more distributed energy production. In the present time we can see that, many houses have a solar panel. Also, smart meter is a main component of the Smart grid. Therefore, it is well understood that, *smart metering* is nothing but a pre-condition for empowering intelligent communication, intelligent controller and smart monitoring in grid applications. Therefore, the fact is that, if we use the IoT platforms in Smart Metering, it will offer us following advantages:

1. The smart meters with an efficient network, empowers faster outage detection and re-establishment of service. A simple example can be, abilities rebound to the benefit of consumers.
2. All consumers always want to have lower bill and controlling of energy or water etc. So IoT platform in smart metering offers consumers, with greater power to controlling their energy or water consumption. Furthermore, it can offer them more selections for handling their bills.
3. IoT platforms in Smart Metering are anticipated to decrease the requirement of constructing power plants. Construction of power plants will be only very much needed, when we will have very expensive occasional peak demand. A more cost-effective method is to figure-out the demand, by either to give incentive to the consumers to decrease their demand, through rates depend on time or other programs. Furthermore, it can be by service-level contracts that permit switching off devices temporarily, which are not required.

3 Security Focuses

3.1 Security Aspects of Existing Projects

In the past work, we have a work that represented an *intelligent Service Security Application Protocol*. It puts together cross-platform communications with authentication, signature, and encryption, to increase IoT apps development abilities. The first fully applied *two-way authentication security scheme* for IoT [54], was described also. It was the Datagram Transport Layer Security (DTLS) protocol, depends on RSA and it is developed for IPv6 over Low Power Wireless Personal Area Networks (6LoWPANs) [6]. Typically, it is positioned in between transport and app layer. It offers message authenticity, confidentiality, and integrity. Some authors have categorized the *Key Management System (KMS)* protocols in four main groups*: key pool framework, mathematical framework, negotiation framework, and public key frame-*

work. The combinatorics-based KMS protocols have some issues, like connectivity and scalability, along with these authentications also.

Now we can have another two appropriate KMS protocols for IoT environs, like Blom and the polynomial schema. In these schemes, numerous counter-measures are obligatory, to be able to have authentication and MitM attacks (man-in-the-middle attack). A framework for IoT depends on Public Key Infrastructure (PKI) is also there. A *transmission model with signature-encryption schemes* is also available now. It deals with the issues like the IoT security requirements (anonymity, trustworthy and attack resistance) with use of Object Naming Service (ONS) queries. It offers some key requirements, data integrity, platform creditability and identities authentication. As we know confidentiality is a major requirement, it was projected with a model with a unique and well resolution, which can guarantee the confidentiality. But this *confidentiality* in the IoT context is still not so good. Moreover, some good efforts have been experimented in the Wireless sensor network (WSN) field. An *authentication protocol* by use of lightweight [18, 35] encryption based on XOR was proposed also. The main idea is manipulation for anti-counterfeiting and privacy protection, coped with constrains IoT devices. A *user authentication* and *key agreement scheme* [19, 20] for Wireless sensor network (WSN) was proposed also. The main idea is to use hash and XOR computations and this technique safeguards, mutual authentication amid gateway nodes (GWN), users and sensor nodes. The authentication and access control method were projected [19, 20]. The idea is to establish a session key on a lightweight [18, 35] encryption mechanism, Elliptic Curve Cryptography (ECC). This scheme explains access control policies based on attribute and it is controlled by an attribute authority, to increase authentication. In IoT, access control denotes to the approvals in the usage of resources, allocated to diverse actors of a wide IoT network. We can see that, these recognized two subjects: *data holders*—feed data accumulators with an exact target, and *data collectors*—identifying and authenticating users and things from the information, which have been acquired.

Some authors [19, 54] have highlighted on the layer, liable for data gathering and also described a categorized access control scheme for the highlighted layer. It offers a single key and needed keys [19, 54] by making use of a deterministic key derivation algorithm. It helps to boost the security and dropping nodes storage costs. Also we got a system depends on an identity, for identifying personal location in emergency circumstances. It comprises of registration, users authentication, policy, and client subsystems [19, 54].

The EU FP7 IoT Work project [23] has projected the Capability Based Access Control (Cap-BAC), which is in use to control the access control processes to services and info with least-privilege operations. The others work is with identity concerns of exact identity management structure for IoT, about authentication and access control in the IoT framework (it was projected an authorization scheme for constrained devices- this scheme actually, integrate Physical Unclonable Functions (PUFs) with Embedded Subscriber Identity Module (eSIM)). This Physical Unclonable Functions (PUFs) based work offers us, authentication, scalability, interoperability, tamper-proof secret keys, compliance with security protocols and it is cheap, secure. Also, one author has described a method for multicast communication secured by use of a common secret key [19, 54] referred as group key. It decreases overhead, network traffic. The highlighting points are, the Protocol can be applied in (1) secure data aggregation in IoT and (2) Vehicle-to-Vehicle (V2V) communications in Vehicular Ad hoc Networks (VANETs).

3.2 Privacy and Trust in IoT

Now let's discuss the current scenario of *privacy* in IoT [1–19]. Some techniques [21–32, 34–40, 52–60] are there for Data tagging, techniques for the Information Flow Control. This technique has contributed for managed privacy [3, 42–44, 54, 56, 58], and it empowers the system to reserve privacy of individuals. Another one is the *User-controlled privacy-preserved access control protocol.* It is based on context-aware k-anonymity privacy policies, privacy defense mechanisms [3, 42–44, 54, 56, 58]. A new approach is called, *continuously Anonymizing STreaming, data* via *adaptive cLustEring (CASTLE).* It is actually a cluster-based scheme and it takes care about, delay constraints of data streams, freshness, enhance privacy preserve techniques and anonymity. The Privacy mechanism can be, like *Discretionary Access* and *Limited Access.* The minimum privacy risk is addressed by this work and also it has a mechanism, to protect the disclosure or cloning of data and avoid attacks. Another idea is the *Privacy protection enhanced DNS (Domain Name System).* It is capable to analyze the privacy risks. This scheme is able to offer the identity authentication and it can protect illegitimate access.

Attribute-based Signature (ABS) scheme, ePASS [57] was also projected. They classify the Attribute based encryption (ABE) into two parts- *Key Policy ABE* and *Cipher-text Policy ABE.* Hence, this is actually a public key encryption scheme, empowers a fine-gained access control, flexible data distribution and scalable key management. It promises *privacy in IoT*, offers attribute privacy for the signer. Another idea is Key-changed mutual authentication protocol [54], for Wireless sensor network (WSN) and Radio-frequency identification (RFID) systems. This protocol integrated a *random number generator* and a *one-way hash function*, reduces risks of replay, replication, DOS, spoofing, and tag tracking.

Trust in IoT [31, 32] can be achieved in different ways. Trust assessments have been carried out in many areas- like—Social networking, Fuzzy technique, Cooperative approach and Identity-based method.

Enforcement in IoT is a big issue. One idea is to deal with various issues like- *network security, security policies, policy enforcement,* and *firewall policy management system.* This idea has to project to use, various security services, like *protecting data confidentiality, integrity, antivirus software, firewalls, authentication, encryption,* and *availability*. Another idea is Policy enforcement languages. This work actually targeted at *uniting policy enforcement* and *analysis languages* and as an outcome it offers, correct policies. Another idea is Web Service Policy (WSP) and eXtensible Access Control Markup Language (XACML). This idea implemented a simulation environment: Web Ontology Language (OWL). It makes use of both policy languages and enforcement mechanisms. Another idea is the Hierarchical Policy Language, for Distributed System (HiPoLDS). It shows us, policy enforcement in distributed reference monitors and how it can manage the flow of info. Another approach can be the enforcement of privacy issues in E-commerce applications. In these paradigms, the approach defends *user anonymity, user trustworthiness* and *customer privacy*. Another idea is a formal and modular framework. This framework can have lots of good features, like it permits to enforce security strategy on concurrent system, generates fault negative and positive. Another idea can be use of algebra *for Communication Process (ACP)* and *Basic Process Algebra (BPA) language*. This idea is able to monitor the requests. Moreover, it can show the satisfaction of correlated rules with an enforcement operator. Another idea is *Access control framework* and *Policy Machine (PM)*. This idea is nothing but integration of enforces policy objectives, expresses and secure framework. But it can be attacked by *Trojan attacks*.

Another idea is about *Discretionary Access Control (DAC) Models, Mandatory Access Control (MAC)* and *Chinese Wall Security Policy Model*. This idea shows us that, Policy Model (PM) is able to impose policy aims. Another approach is about, semantic web framework and meta-control model. Their work is combining the policy reasoning with identification and right of entry to the sources of information. Another idea is an enforcement resolution, denoted as SecKit. The SeeKit is depend on Model-based Security Toolkit. The work is, integrated with *MQ Telemetry Transport (MQTT) protocol layer*. It also guarantees enforcement of security and privacy policies. Another idea is the VIRTUS Middleware, which is able to offer, reliable and secure communication channel for distributed apps. It has an *eXtensible Messaging and Presence Protocol (XMPP)*.

Another idea is Aml Framework and Otsopack. It can run on various platforms (Java SE, Android) and it's extensible, modular and simple. Another idea is *Trivial File Transfer Protocol (TFTP)*. It has more and more techniques for increasing the trust, privacy, and security in embedded system infrastructures. *Naming, Addressing and Profile Server (NAPS)* is another idea. It can work as a main module at the back-end data centre, to downstream from apps content-based data filtering, matching and support the upstream. Another idea is about A security architecture for IoT transparent middleware. The main contribution is that, it is depend on the existing security approaches (AES, TLS and oAuth). The architecture offers, a perfect blend of confidentiality, authenticity, integrity and privacy. Another idea is *Heterogeneity Inclusion and Mobility Adaptation through Locator ID Separation (HIMALIS)*, which has three much related main contributions- privacy vulnerabilities, supports scalable inter-domain authentication, and solves security issue.

Another idea is an *Ultra-lightweight and privacy preserving authentication protocol*, which offers privacy and protects from many attacks. Another idea is the *Mobile Intrusion Prevention System (m-IPS)*. It offers specific access control. Another idea is *Mobile Sensor Data Processing Engine (MOSDEN)*. It empowers to acquire, and process sensor data and it works fine with push and pull data streaming mechanisms.

Also, NSF, National Science Foundation—has multi-institutional projects [28] on security with more highlight on Cyber physical security. They are trying to find ideal network architectures and networking with *very high efficiency*. Moreover, they are working with new kind of communication protocols with features like trusted data, trusted models, integrity and authentication. They are also trying to resolve issues like use of network assets on mobile environs, technical challenges, and the trade-offs of among mobility and scalability.

FIRE, Future Internet Research and Experimentation [29, 30] is a multi-nation project of EU, China and Korea. They are exploring several resolutions for the *positioning of IoT technologies in numerous application areas*. Some of the prominent application areas are people livelihood, medical and health service, urban management, public safety and social security. Their major concerning areas are intellectual property right, information security and privacy.

EU with Japan have launched an ICT Cooperation project [22]. They are trying to offer *common global standards*. These common global standards have firm target, to confirm an all-in-one communications and common ways to store and have right to use the information. These global standards can ensure us the highest security and energy efficiency standards.

4 Our Approach

What we have found in the above projects discussed, some have efficiency but not more secure, then some have just some sort of security, which can be broken very easily. Some projects have privacy but not proper authentications. Similarly, End to End secure communication is missing in some of those projects. Moreover, some projects security is good in security aspects but it's not lightweight, so we know for IoT and IoE, we need lightweight cipher. We *need a cipher for End to End encrypted communication*, which can be much more secure and more authenticated. Furthermore, very popular public-key cryptosystem RSA (made of the initial letters of the surnames of Ron Rivest, Adi Shamir, and Leonard Adleman) has many backlogs like.

1. *Exploitation* of multiplicative property and *exploitation* of Homomorphic property.
2. *Difficulty* of the integer factorization problem and Computational modular exponentiation *complexity* problem.
3. Partial key exposure *vulnerability* and Low Modular *complexity* with *Effortlessness* and *speediness* problem.
4. Real-time Key negotiation between each peers' problem and parallel protection to Sniffing attacks.
5. Chosen Cipher text attacks (CCA), Brute force key search, and Timing attacks.
6. Asymptotic very *low speed of decryption* etc.

We have developed a *Secure Hybrid RSA (SHRSA) messaging system* for End to End encrypted messaging [46–51] with solutions to many bottlenecks of RSA and with high efficiency and lightweight architecture. We first took some of the RSA

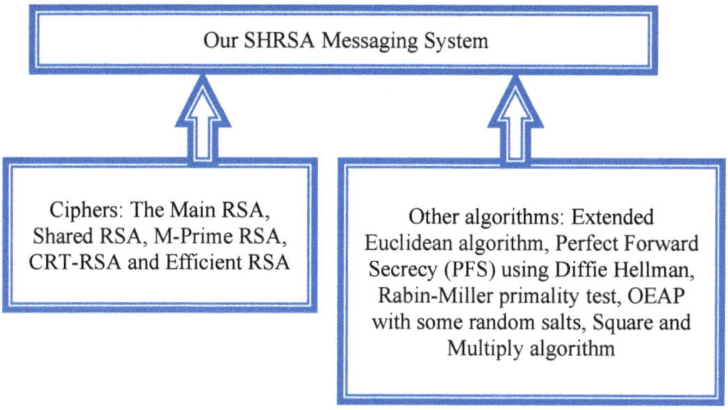

Fig. 3 Our SHRSA Messaging system scheme

variants and insert them in our SHRSA encryption and decryption [46–51], with some other algorithms, to resolve some of the major problems with main RSA, as shown in Fig. 3. As an outcome, we have developed a SHRSA messaging scheme with SHRSA End to End Encryption and SHRSA Decryption.

Instant Messaging (IMs) schemes nowadays have many backlogs. Some of the Instant Messaging (IMs) schemes' backlogs are-

1. Centralized system, so single point failure can occur anytime, anyplace.
2. Only messages are encrypted, not strong encryption for communicating party's communication protocol.
3. Decryption is not faster. Statistical complexity is less and vulnerable to chosen text cipher attacks and other attacks.
4. Authentication is by only password.
5. The requirement of a third party is considered a disadvantage and even when a third-party is present, it is often considered as a disinterested party. Nowadays lots of Instant Messaging (IMs) schemes have third party security.
6. Insecure default settings on Instant Messaging (IMs) schemes for clients are a big problem.
7. *Sharing* Instant Messaging (IMs) features with other applications introduce significant security risks.
8. Impersonation using a stolen/compromised password cannot generally be prevented in password-only systems, as a *password* is the only piece of secret shared between a user and the IM server.
9. Denial of Service (DoS) attack is a big problem.
10. Pure *Peer to Peer scheme* is used in very less cases.
11. Using *Secure Sockets Layer (SSL)*-based solutions for public IM service has drawbacks, while it is a step forward in terms of security.
12. The use of unpublished, non-standard proprietary *protocols* and non-centralized *peer-to-peer file transfer* makes it difficult to monitor IM traffic.

13. Almost all popular Instant Messaging (IMs) connections lack *authentication* (except in the login message), *confidentiality* and *integrity*. This opens the door to many other *security vulnerabilities* including impersonation, denial of service (DoS), man-in-the-middle, replay, etc.

Our SHRSA messaging system [46–51] will replace these following disadvantages of existing Instant Messaging schemes and protocols which are in use now-

1. Its *distributed*, no single point failure with SHRSA and its peer to peer connection, works with n*n SHRSA servers and clients.
2. SHRSA Decryption is 9 time faster.
3. SHRSA encryption is much more complex between each peer.
4. PH (Pohlig-Hellman) key exchange and Diffie-Hellman Exchange key ensure three-way authentications peer to peer.
5. Optimal Asymmetric Encryption Padding (OAEP) with some random salts added on runtime with synchronize time gap in our SHRSA scheme, protects Chosen Cipher Text Attack and short plaintext attack, Man-in the Middle attack and other attacks.
6. SHRSA works with End to end encryption with full mesh topology.
7. No SSL used and also no external digital certificates are used, we have our own SHRSA's complex security, authentication and very strong confidentiality.
8. No default settings are shared with others, so less vulnerable.
9. No need of any third-party security, so cost saving.
10. It's more reliable, more efficient and stronger due to variants of RSA integration.
11. No need to install IMSecure.
12. No need of use of any password as we have our own three-way four layers authentications for peers and then SHRSA encryption.

Our SHRSA messaging system [46–51], works with an End to End encryption model with Full Mesh networked architecture to ensure pure *peer to peer* nature. We have designed our SHRSA messaging scheme, with nine layers protocol stack as shown in Fig. 4.

Moreover, in our past [46–51] work, we have shown that our SHRSA is a perfect combination of strong security, authentication, and reliability. In the encryption level, our *SHRSA encryption* with 1024 Bit RSA modulus, is helping us to resolve some of the scientific problems like,

(a) The exploitation of multiplicative property.
(b) The exploitation of homomorphic property (meet-in-the-middle attack).
(c) Difficulty of the integer factorization problem of RSA.
(d) The very high computationally costly exponentiation modulo N problem.
(e) Low modular complexity with effortlessness and speediness problem.

Moreover, our SHRSA encryption scheme, have proper protection from *Chosen Plaintext Attack* and *short plaintext attack* etc., along with protection to *Sniffing attacks* and *resolving the real-time Key negotiation issue* also. *Brute force attack*

```
+--------------------------------------------------------------------------------+
| SHRSA Messaging Decryption Layer |
+--------------------------------------------------------------------------------+
| SHRSA Messaging Encryption Layer |
+--------------------------------------------------------------------------------+
| E2E Secure Channel Layer|
+--------------------------------------------------------------------------------+
| RSA-OAEP layer |
+--------------------------------------------------------------------------------+
| Perfect Forward Secrecy (PFS) based on Diffie-Hellman (DH) |
+--------------------------------------------------------------------------------+
| Peer to Peer Authentication by Diffie-Hellman Key exchange |
+--------------------------------------------------------------------------------+
| 3-way Handshaking |
+--------------------------------------------------------------------------------+
| Diffie-Hellman Key exchange |
+--------------------------------------------------------------------------------+
| E2E Connection Layer |
+--------------------------------------------------------------------------------+
```

Fig. 4 Our SHRSA nine-layer protocol stack

is countered by randomly altering the keys in synchronous time slot with 1024 Bit value.

In the decryption level of SHRSA, our SHRSA decryption is helping us to resolve some of the scientific problems like,

(a) Computational modular exponentiation complexity.
(b) Partial key exposure vulnerability.
(c) Asymptotic very low speed of decryption of RSA problem. We are gaining almost 9 times faster asymptotic decryption speed.

Cost estimation comparisons of variants of RSA, CRT-RSA and our SHRSA system's decryption is here Fig. 4. (where $k = 3$(no. of primes)). We have found that SHRSA scheme's decryption time is near about *290* ms (average running of decryption class 5 times of RSA, CRT-RSA and SHRSA APIs for decryption during messaging), whereas CRT-RSA decryption has taken time near about *652* ms and RSA decryption has taken 2611 ms. But all cipher's *encryption time* is same (Fig. 5).

Hence, it iss a complex secure, efficient and lightweight system for use it in present IoT and in near future in Future Internet of everything (IoE), though we have used our SHRSA cipher for secure and efficient messaging scheme as on time [46–51]. It's a distributed system, so no chance of central failure without depending upon third party for authentication and security. As it's implemented in Java, so it's interoperable also.

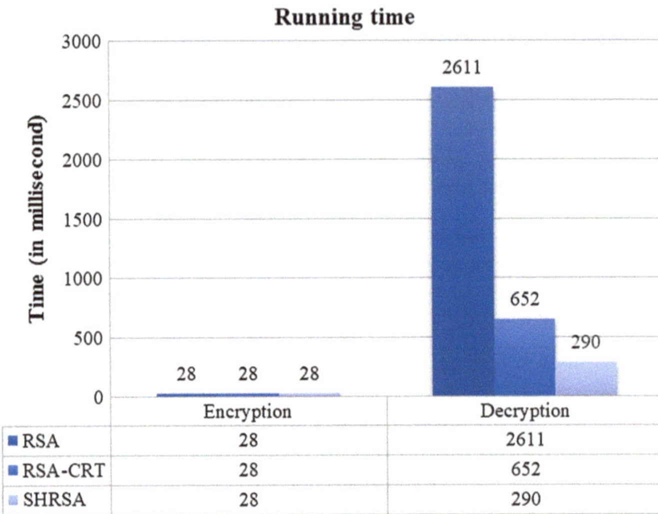

Fig. 5 Speed-up comparison of RSA variants and SHRSA

5 Conclusions

To have a proper secure and privacy protected proliferation of IoT services, we need architectures with ciphers or other security approaches to entail customized security and privacy levels. In this paper we have discussed several existing models of IoT, it has given us a wide-range overview of many open issues with future directions in the IoT security field. We have discussed various issues here like, trust, privacy and security rules in the middleware environs and for mobile devices, diverse technologies and communication standards, the security and privacy necessities and appropriate security resolutions. In precise, the secured IoT necessitate compliance with well-defined security and privacy strategies, privacy for users and things, confidentiality, access control, and trustworthiness among devices and users. We also have described our SHRSA messaging scheme with 9-layered protocol stack, which has many real-time applications and it is ready for use as our system is installable software now. Our SHRSA messaging scheme's encryption and decryption have not *only replaced many bottlenecks* of popular cipher RSA but also has resolved *many problems of existing Instant Messaging (IM) schemes*. In the real-time testing results, we have found that SHRSA scheme's decryption time is near about *290* ms (average running of decryption class 5 times of RSA, CRT-RSA and SHRSA APIs for decryption during messaging), whereas CRT-RSA decryption has taken time near about *652* ms and RSA decryption has taken 2611 ms. Here we have gained practically *9 times in decryption* by our SHRSA than RSA. But RSA, CRT-RSA and SHRSA's *encryption time* is same. Due to multiple cipher integration it has already strong *security, authentication* and *privacy*. Our implementation has allowed ubiquitous and automatic encryption available to all users without any need of understanding the complications involved. Our architecture also affords a hassle-free, secure, peer-to-peer, unconventionally strong and reliable platform with End to End -encryption for people and organizations who are concerned about their privacy and security. Future researches in the security concerns of the Internet of Things would mostly quintessence on the consequential characteristics like: terminal security function, related laws for the security of the Internet of Things, the open security system, single privacy protection mode, etc. It is unconditional that, the security of the Internet of Things is more than a technical difficulty, which also has necessities for series of policies, laws and regulations, perfect security management system for mutual collocation.

Acknowledgements This research is supported by National Natural Science Foundation of China (No. 61631013). We want to convey our gratitude and tribute to Late Prof. Wang Jing for his constant supervision and encouragement for this project.

References

1. Jara, A., Kafle, V., Skarmeta, A.: Secure and scalable mobility management scheme for the internet of things integration in the future internet architecture. Int. J. Ad Hoc Ubiquitous Comput. **13**(3–4), 228–242 (2013)
2. Li, S., Gong, P., Yang, Q., Li, M., Kong, J., Li, P.: A secure handshake scheme for mobile-hierarchy city intelligent transportation system. In: International Conference on Ubiquitous and Future Networks. ICUFN, Da Nang, pp. 190–191 (2013)
3. Kang, K.C., Pang, Z.B., Wang, C.C.: Security and privacy mechanism for health internet of things. J. China Univ. Posts Telecommun. **20**(Suppl 2), 64–68 (2013)
4. Goncalves, F., Macedo, J., Nicolau, M., Santos, A.: Security architecture for mobile e-health applications in medication control. In: 2013 21st International Conference on Software, Telecommunications and Computer Networks. SoftCOM, Primosten, pp. 1–8 (2013)
5. An, J., Gui, X., Zhang, W., Jiang, J., Yang, J.: Research on social relations cognitive model of mobile nodes in internet of things. J. Netw. Comput. Appl. **36**(2), 799–810 (2013)
6. Kasinathan, P., Costamagna, G., Khaleel, H., Pastrone, C., Spirito, M.: Demo: an ids Framework for Internet of Things Empowered by 6lowpan, Berlin, Germany, pp 1337–1339 (2013)
7. BETaaS Consortium: BETaaS building the environment for the things as a service D2. 2. 2–Specification of the extended capabilities of the platform, pp. 1–61 (2014)
8. IoT-A Consortium (2014) IoT-A unified requirements. http://www.iot-a.eu/public/requirements/.31Jan2014
9. Gao, L., Bai, X.: A unified perspective on the factors influencing consumer acceptance of internet of things technology. Asia Pac. J. Mark Logist. **26**(2), 211–231 1075 (2014)
10. Gazis, V.: Carlos Garcia Cordero, Emmanouil Vasilomanolakis, Panayotis Kikiras, and Alex Wiesmaier. Security perspectives for collaborative data acquisition in the internet of things. In: International Conference on Safety and Security in Internet of Things. Springer, New York 1079 (2014)
11. IoT-A Consortium (2014) IoT-A—Internet of things architecture. http://www.iot-a.eu/. 27 Jan 2014
12. Logvinov, O., Kraemer, B., Adams, C., Heiles, J., Stuebing, G.: Mary Lynne Nielsen, and Brenda Mancuso. Standard for an architectural framework for the internet of things (IoT) IEEE P2413 Webinar Panelists, pp. 1–12 (2014)
13. Zanella, A., Bui, N., Castellani, A.P., Vangelista, L., Zorzi, M.: Internet of things for smart cities. IEEE Internet Things J. **1**, 22–32 (2014)

14. Grieco, L.A., Alaya, M.B., Monteil, T., Drira, K.K.: Architecting information centric ETSI-M2M systems. In: IEEE PerCom (2014)
15. Anderson, J., Rainie, L.: The internet of things will thrive by 2025, Pew research internet project (2014). http://www.pewinternet.org/2014/05/14/internet-of-things/
16. Yan, Z., Zhang, P., Vasilakos, A.V.: A survey on trust management for internet of things. J. Netw. Comput. Appl. **42**, 120–134 (2014)
17. Piro, G., Boggia, G., Grieco, L.A.: A standard compliant security framework for IEEE 802.15.4 networks. In: Proceedings of IEEE World Forum on Internet of Things (WF-IoT), Seoul, South Korea, pp. 27–30 (2014)
18. Lee, J.-Y., Lin, W.-C., Huang, Y.-H.: A lightweight authentication protocol for internet of things. In: 2014 International Symposium on Next-Generation Electronics, ISNE 2014, Kwei-Shan, pp. 1–2 (2014)
19. Turkanovi, M., Brumen, B., Hlbl, M.: A novel user authentication and key agreement scheme for heterogeneous ad hoc wireless sensor networks, based on the internet of things notion. Ad Hoc Netw. **20**, 96–112 (2014)
20. Ye, N., Zhu, Y., Wang, R.-C.B., Malekian, R., Lin, Q.-M.: An efficient authentication and access control scheme for perception layer of internet of things. Appl. Math. Inf. Sci. **8**(4), 1617–1624 (2014)
21. BUTLER Project. http://www.iot-butler.eu
22. EU-Japan Project. http://www.eurojapan-ict.org/
23. European FP7 IoT@Work project. http://iot-at-work.eu
24. HYDRA Project. http://www.hydramiddleware.eu/
25. Usable Trust in the Internet of Things. http://www.utrustit.eu/
26. iCORE Project. http://www.iot-icore.eu
27. HACMS Project. http://www.defenseone.com/technology
28. National Science Foundation Project. http://www.nsf.gov
29. FIRE EU-China Project. http://www.euchina-fire.eu/
30. FIRE EU-Korea Project. http://eukorea-fire.eu/
31. Gu, L., Wang, J., Sun, B.B.: Trust management mechanism for internet of things. China Commun. **11**(2), 148–156 (2014)
32. Liu, Y.-B., Gong, X.-H., Feng, Y.-F.: Trust systembased on node behavior detection in internet of things. Tongxin Xuebao/J Commun. **35**(5), 8–15 (2014)
33. Singh, J., Bacon, J., Eyers, D.: Policy enforcement within emerging distributed, event-based systems. In: DEBS 2014—Proceedings of the 8th ACM International Conference on Distributed Event-Based Systems, pp. 246–255 (2014)
34. Neisse, R., Steri, G., Baldini, G.: Enforcement of security policy rules for the internet of things. In: Proceedings of IEEE WiMob, Larnaca, Cyprus, pp. 120–127 (2014)
35. Gòmez-Goiri, A., Orduna, P., Diego, J., de Ipina, D.L.: Otsopack: lightweight framework for interoperable ambient intelligence applications. Comput. Hum. Behav. **30**, 460–467 (2014)
36. Wang, Y., Qiao, M., Tang, H., Pei, H.: Middleware development method for internet of things. Liaoning Gongcheng Jishu Daxue Xuebao (Ziran Kexue Ban)/J Liaoning Tech Univ (Nat Sci Ed) **33**(5), 675–678 (2014)
37. Ferreira, H., De Sousa, R. Jr, De Deus, F., Canedo, E.: Proposal of a secure, deployable and transparent middleware for internet of things. In: Iberian conference on Information Systems and Technologies. CISTI, Barcelona, pp. 1–4 (2014)
38. Niu, B., Zhu, X., Chi, H., Li, H.: Privacy and authentication protocol for mobile RFID systems. Wirel. Pers. Commun. **77**(3), 1713–1731 (2014)
39. Jeong, Y.-S., Lee, J., Lee, J.-B., Jung, J.-J., Park, J.: An efficient and secure m-IPS scheme of mobile devices for human-centric computing. J. Appl. Math. **2014**, 1–8 (2014)
40. Geng, J., Xiong, X.: Research on mobile information access based on internet of things. Appl. Mech. Mater. **539**, 460–463 (2014)
41. Kubler, S., Frmling, K., Buda, A.: A standardized approach to deal with firewall and mobility policies in the IoT. Pervasive Mobile Comput. **20**, 100–114 (2014)

42. Daubert, J., Wiesmaier, A., Kikiras, P.: A view on privacy & trust in IoT. In: IOT/CPS-Security Workshop, IEEE International Conference on Communications, ICC 2015, London, GB, June 08–12, 2015, page to appear. IEEE (2015)
43. Sadeghi, A.R., Wachsmann, C., Waidner, M.: Security and privacy challenges in industrial internet of things. In: Annual Design Automation Conference, p. 54. ACM, New York (2015)
44. Sicari, S., Rizzardi, A., Grieco, L.A., Coen-Porisini, A.: Security, privacy and trust in internet of things: the road ahead. Comput. Netw. **76**, 146–164 (2015)
45. Zhang, Z.-K., Cheng, M., Cho, Y., Shieh, S.: Emerging security threats and countermeasures in IoT. In: ACM Symposium on Information, Computer and Communications Security, pp. 1–6. ACM, New York (2015)
46. Bhattacharjya, A., Zhong, X., Wang, J.: Strong, efficient and reliable personal messaging peer to peer architecture based on Hybrid RSA. In: Proceedings of the International Conference on Internet of Things and Cloud Computing (ICC 2016) ISBN 978-1-4503-4063-2/16/03. The Møller Centre-Churchill College, Cambridge (2016). https://doi.org/10.1145/2896387.2896431
47. Bhattacharjya, A., Zhong, X., Wang, J.: An end to end users two way authenticated double encrypted messaging scheme based on Hybrid RSA for the future internet architectures. Int. J. Info Comput. Secur. **10**, 63–79 (2017). https://doi.org/10.1504/IJICS.2018.10005506
48. Bhattacharjya, A., Zhong, X., Wang, J., et. al.: On Mapping of address and port using translation (MAP-T). Int. J. Info Comput. Secur. **11**(3), 214–232 (2019)
49. Bhattacharjya A., Zhong X., Wang J.: HYBRID RSA based highly efficient, reliable and strong personal Full Mesh Networked messaging scheme. Int. J. Info Comput. Secur. **10**(4), 418–436 (2018)
50. Bhattacharjya A., Zhong X., Wang J., et al.: Security challenges and concerns of internet of things (IoT), cyber-physical systems: Architecture, security and application. EAI/Springer Innovations in Communication and Computing, pp 153–185 (2019)
51. Bhattacharjya A., Zhong X., Wang J., et al.: Secure IoT structural design for smart homes, smart cities cybersecurity and privacy, Elsevier, pp 187–201 (2019). http://www.sciencedirect.com/science/article/pii/B9780128150320000135
52. Cherkaoui, A., Bossuet, L., Seitz, L., Selander, G., Borgaonkar, R.: New paradigms for access control in constrained environments. In: 2014 9th International Symposium on Reconfigurable and Communication-Centric Systems-on-Chip (ReCoSoC), Montpellier, pp 1–4 (2014)
53. Tormo, G.D., Marmol, F.G., Perez, G.M.: Dynamic and flexible selection of a reputation mechanism for heterogeneous environments. Futur Gener. Comput. Syst. **49**, 113–124 (2014)
54. Peng, L.B., Ru-chuan, W.B., Xiao-yu, S., Long, C.: Privacy protection based on key-changed mutual authentication protocol in internet of things. Commun. Comput. Inf. Sci. **418**, 345–355 (2014)
55. Sicari, S., Rizzardi, A., Cappiello, C., Coen-Porisini, A.: A NFP model for internet of things applications. In: Proceedings of IEEE WiMob, Larnaca, Cyprus, pp. 164–171 (2014)
56. Wang, X., Zhang, J., Schooler, E., Ion, M.: Performance evaluation of attribute-based encryption: toward data privacy in the IoT. In: 2014 IEEE International Conference on Communications, ICC 2014, Sydney, NSW, pp. 725–730 (2014)
57. Su, J., Cao, D., Zhao, B., Wang, X., You, I.: ePASS: an expressive attribute-based signature scheme with privacy and an unforgeability guarantee for the internet of things. Futur. Gener. Comput. Syst. **33**, 11–18 (2014)
58. Ukil, A., Bandyopadhyay, S., Pal, A.: IoT-privacy: to be private or not to be private. In: Proceedings—IEEE INFOCOM, Toronto, ON, pp. 123–124 (2014)
59. Sicari, S., Cappiello, C., Pellegrini, F.D., Miorandi, D., Coen-Porisini, A.: A security-and quality-aware system architecture for internet of things. Inf. Syst. Front. **18**, 1–13 (2014)
60. Colistra, G., Pilloni, V., Atzori, L.: The problem of task allocation in the internet of things and the consensus-based approach. Comput. Netw. **73**, 98–111 (2014)

Aniruddha Bhattacharjya is with the Department of Electronic Engineering, Tsinghua University, Beijing, China, as a Chinese Government Ph.D. scholar. His research interests are cryptography, Network security, RFID-based architectures and middleware, security in fixed and wireless Networks, applications of cryptography, and IoT security. He has received the ICDCN 2010, Ph.D. Forum Fellowship. He achieved the best paper award in ACM ICC 2016, in Cambridge University, UK. Since 2012, he has been working as an IEEE mentor and ACM faculty sponsor. He is a member of 34 IEEE societies and various IEEE technical committees. He has published 35 papers as well as three pending US patents and one Chinese innovation patent.

Xiaofeng Zhong received his Ph.D. in Information and Communication Systems from Tsinghua University in 2005. He is an Associate Professor in the Department of Electronic Engineering at Tsinghua University. He performs research in the field of mobile networks, including users' behaviors and traffic model analyses, MAC and network protocol design, and resource management optimization. He has published more than 30 papers and holds seven patents.

Jing Wang received his BS and MS degree in Electronic Engineering from Tsinghua University, Beijing, China in 1983 and 1986, respectively. He has worked as a Faculty member in Tsinghua University since 1986. He is currently a Professor at the School of Information Science and Technology. He also serves as the Vice Director of the Tsinghua National Laboratory for Information Science and Technology. His research interests are in the area of wireless communications, including transmission and networking technologies of 4G/5G. He has published more than 150 conference and journal papers.

Xing Li received the B.S. degree in radio electronics from Tsinghua University, Beijing, China, in 1982, and the M.S. and Ph.D. degrees in electrical engineering from Drexel University, Philadelphia, PA, USA, in 1985 and 1989, respectively. He is currently a Professor with the Electronic Engineering Department, Tsinghua University. His research activities and interests include statistical signal processing, multimedia communication, and computer networks. He has published more than 300 papers in his research areas. He is a Deputy Director of the China Education and Research Network (CERNET) Center and a Member of the Technical Board of the CERNET Project. He was a Member of Communication Expert Committee of the China National "863" High Technology Project. He is a Formal Chairman of the Asia Pacific Networking Group and a Formal Member of the executive council of the Asia Pacific Network Information Center.

IoT Security, Privacy, Safety and Ethics

Hany F. Atlam and Gary B. Wills

Abstract The Internet of Things (IoT) represents a revolution of the Internet which can connect nearly all environment devices over the Internet to share their data to create novel services and applications for improving our quality of life. Using cheap sensors, the IoT enables various devices and objects around us to be addressable, recognizable and locatable. Although the IoT brought infinite benefits, it creates several challenges, especially in security and privacy. Handling these issues and ensuring security and privacy for IoT products and services must be a fundamental priority. Users need to trust IoT devices and related services are secure. Moreover, the IoT safety must be considered to prevent the IoT system and its components from causing an unacceptable risk of injury or physical damage and at the same time considering social behaviour and ethical use of IoT technologies to enable effective security and safety. This chapter provides a discussion of IoT security, privacy, safety and ethics. It starts by providing an overview of the IoT system, its architecture and essential characteristics. This is followed by discussing IoT security challenges, requirements and best practices to protect IoT devices. The IoT privacy is also discussed by highlighting various IoT privacy threats and solutions to preserve the privacy of IoT devices. The IoT safety, ethics, the need for the ethical design and challenges encountered are also discussed. In the end, smart cities are introduced as a case study to investigate various security threats and suggested solutions to maintain a good security level in a smart city.

Keywords Internet of things · IoT security and privacy · Ethical design for IoT · IoT safety · Security by design · Privacy by design

H. F. Atlam (✉) · G. B. Wills
School of Electronics and Computer Science, University of Southampton, Southampton, UK
e-mail: hfa1g15@soton.ac.uk

G. B. Wills
e-mail: gbw@soton.ac.uk

H. F. Atlam
Department of Computer Science and Engineering, Faculty of Electronic Engineering, Menoufia University, Menoufia, Egypt

M. Farsi et al. (eds.), *Digital Twin Technologies and Smart Cities*, Internet of Things, https://doi.org/10.1007/978-3-030-18732-3_8

1 Introduction

Currently, the Internet of Things (IoT) has become one of the hottest topics among researchers and experts. It is considered a universal presence that allows all objects/things in our environment to be connected over the Internet with the capability to interconnect with each other without human intervention. The IoT involves a variety of objects that can be connected using either wireless or wired networks. These objects have a unique addressing scheme that allows them to interact and cooperate with each other to create novel applications and services such as smart homes, smart transportation, connected cars, smart grids, smart cities, smart traffic control and others [1].

The social acceptance of IoT applications and services is strongly depending on the trustworthiness of information and the protection of private data. Since the IoT is a complex, distributed and heterogeneous system in nature, it faces several challenges regarding security and privacy. Currently, building an effective and reliable security technique is one of the highest priorities to consider [2]. Although a number of researchers have introduced several solutions to the security and privacy issues, a reliable security technique for the IoT is still in demand to satisfy requirements of data confidentiality, integrity, privacy and trust [3].

In addition, the IoT safety is considered to be one of the highest significances to prevent the IoT and its elements from producing physical damage or undesirable threat and protect the surrounding environment from such damage. Building the IoT system with embedded safety and reliability features should be considered to create new design architectures that provide a safe and reliable system environment [4]. In addition, there is a need to develop an ethical framework that ensures the IoT is used for the good of humanity and not the other way around. A strong ethical standard will motivate companies to design smarter and more inclusively products to avoid algorithmic issues and ensure global connectivity.

The main objective of this chapter is to provide an overview of IoT security, privacy, safety and ethics. It starts by discussing the architecture and essential characteristics of the IoT system. This is followed by investigating IoT security by highlighting security requirements, security by design, security attacks and security challenges of the IoT system. IoT privacy with investigating privacy threats and suggested solutions are also discussed. Also, IoT safety and ethics are investigated by highlighting the need for ethical design and ethics challenges in the IoT system. In the end, a case study of the smart city is introduced to discuss security threats and suggested solutions in the smart city context.

The rest of this chapter is structured as follows: Sect. 2 provides an overview of the IoT system; Sect. 3 discusses IoT security; Privacy issues and suggested solutions are discussed in Sect. 4; Sect. 5 discusses IoT safety; the need for ethical design and ethics challenges in the IoT are presented in Sect. 6; Sect. 7 discusses security issues and suggested solution in the smart city context; Sect. 8 is the conclusion.

2 An Overview of IoT

This section provides an overview of the IoT system by discussing IoT definitions, its layer architecture and essential characteristics.

2.1 IoT Definition

The IoT system has evolved to involve the perception of realizing a global infrastructure of interconnected networks of physical and virtual objects. These objects/things are interconnected using either wired or wireless networks to share information between various IoT devices to create novel applications and services [5].

Originally, the notion of the IoT was initially presented by Kevin Ashton, who is the originator of MIT auto-identification centre in 1999 [6]. Ashton has said, '*The Internet of Things has the potential to change the world, just as the Internet did. Maybe even more so*' [6]. Later, the IoT was officially presented by the International Telecommunication Union (ITU) in 2005 [7]. The IoT has been defined by many organizations and researchers. However, the definition provided by ITU in 2012 is the most common. It stated: '*a global infrastructure for the information society, enabling advanced services by interconnecting (physical and virtual) things based on, existing and evolving, interoperable information and communication technologies*' [8].

In addition, Guillemin and Friess [9] have suggested one of the simplest definitions that describe the IoT in a smooth manner, as shown in Fig. 1. It stated: '*The Internet of Things allows people and things to be connected Anytime, Anyplace, with anything and anyone, ideally using any path/network and any service*'. Several definitions were suggested by many researchers describing the IoT system from different perspectives but the important thing that majority of researchers have agreed on the IoT is created to increase information sharing that leads to a better world for all the human beings.

2.2 IoT Essential Characteristics

The IoT represents a promising technology that aims to improve people's quality of life by generating new applications that facilitate people daily activities. For the IoT system, there are a set of common features, which include the following:

- **Large Scale**: IoT devices are increased in billions. This large-scale network of devices needs to be controlled to allow devices to communicate with each other. In addition, this large-scale network generates a huge amount of data which produce a critical issue regarding data interpretation and analysis.
- **Intelligence**: Combining sophisticated software algorithms with hardware allow IoT devices to become smart. These intelligence abilities allow IoT devices to

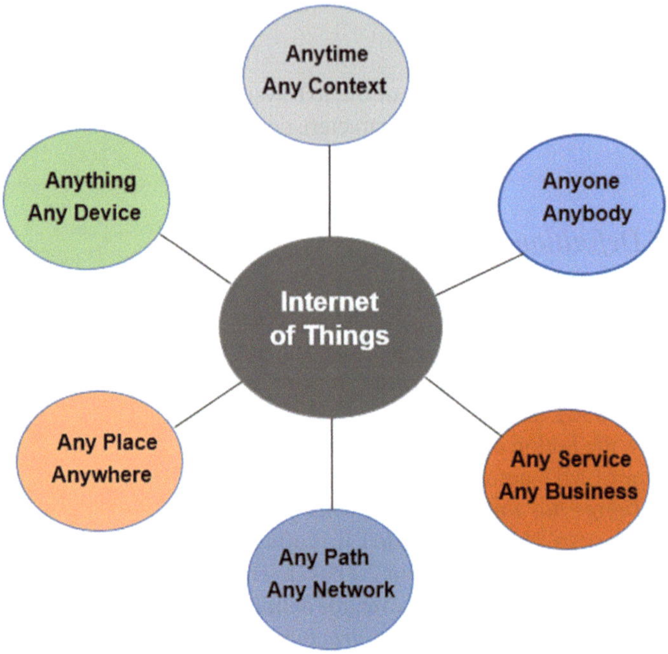

Fig. 1 The IoT can connect anything anywhere using any path

make intelligent decisions in various situations and interact intelligently with other communicating devices.

- **Sensing**: Sensors are the main part of the IoT system which are used to perceive changes in the surrounding environment and create data that reveal their status. With various sensing technologies, sensors provide a good understating of surroundings and increase human awareness about the physical world [8].
- **Complex System**: The IoT system consists of billions of heterogamous objects with different hardware and software capabilities that make the management process a very difficult task to accomplish especially with constraints associated with memory, energy and time.
- **Dynamic Environment**: The IoT has the ability to connect almost all objects of our environment without having to determine the IoT network boundaries which make it a dynamic system in nature. Also, IoT devices can operate and be adjusted dynamically based on changing conditions and situations.
- **Massive Amount of Data**: As IoT devices are in billions. These devices sense their surroundings and generate a huge amount of data which make it one of the sources of what is called Big Data.
- **Heterogeneity**: The IoT system involves billions of devices with heterogeneous features such as operating systems, platforms, communication protocols and others. These heterogeneous features make the management operation a complex task to perform.

- **Limited Energy**: Most IoT devices are small and lightweight with limited resources, so they are designed to work with minimal energy consumption.
- **Connectivity**: One of the main features of the IoT system is the ability to connect various devices with different characteristics and use their shared information to create novel applications and services.
- **Self-configuring**: Devices are configured to perform a certain operation. But for IoT devices, they have the capability of self-configuring that enable them to operate without human intervention. IoT devices could configure themselves to the up-to-date software in association with the device manufacturer without user involvement.
- **Unique Identity**: Within the IoT network, each IoT object is identified and recognized using a unique identifier such as the IP address. These identities are provided by IoT manufactures to use it to upgrade devices to the appropriate platforms. In addition, these devices have interfaces that enable users to collect the required information from the devices, record their status and manage them remotely.
- **Context awareness**: In the IoT environment, there are multiple sensors that sense their surroundings, collect and store the required information, these sensors may take decisions based on collected data which make it a context aware.

2.3 IoT Architecture

IoT World Forum (IWF) architecture committee released an IoT reference model in October 2014 [10]. This model works as a common framework to help the industry to accelerate IoT deployments. This reference model is intended to consolidate and encourage the collaboration and development of IoT deployment models. It is designed as seven layers so that each layer provides additional information for establishing a common terminology, as shown in Fig. 2. It also classifies where various kinds of processing are operated through different layers of the IoT reference model. In addition, this model enables various manufacturers to produce IoT products that are compatible with each other, which convert the IoT from a conceptual model into a real and approachable system. Layer 1 is the physical layer. It contains physical devices and controllers that manage various objects. These objects represent things in the IoT that involve various types of devices that send and receive information, for instance, sensors that collect information about the surrounding environment [11]. Communications and connectivity are in layer 2. This layer is used to interconnect different IoT things with each other using interconnection devices such as switches, gateway, router and firewalls. Layer 3 is edge computing. This layer takes date coming from the connectivity layer and converts it into information appropriate for storage and higher level processing. At this layer, the processing components work with a huge amount of data and it may execute some data transformation to reduce the size of data. Layer 4 is the data accumulation. This layer is concerned with storing data coming from different IoT devices. This data is filtered and processed by the edge-computing layer that absorbs large quantities of data and places them in

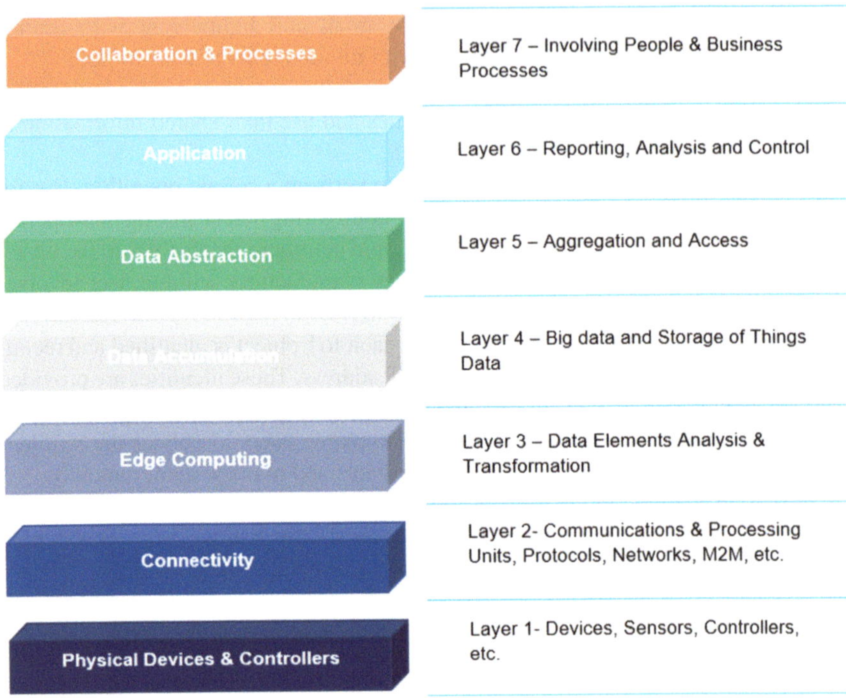

Fig. 2 The IoT reference model according to IWF, drawn from data provided in [13]

storage to be accessible by higher levels. Different types of data in various formats and from heterogeneous processors may come up from the edge-computing layer for storage. The data abstraction layer aggregates and formats stored data in a way that make them accessible by applications in a more manageable and efficient way. Layer 6 is the application layer. This layer is concerned with the information interpretation of various IoT applications. This layer encompasses a variety of applications that use IoT input data or control IoT devices [10]. The collaboration and processes are in layer 7. This layer identifies individuals who can communicate and collaborate to make the IoT system more useful. It also involves various applications to exchange data and control information over the Internet.

3 IoT Security

Majority of researchers and experts have confirmed that securing the IoT system is one of the most serious challenges that stand in the way of successful adoption of IoT devices. The value of the IoT system comes from connecting all small and large systems together and allowing them to communicate with each other over the Internet.

Since the IoT is a dynamic system in nature in which every poorly secured object can disturb the security and resilience of the entire system as they are connected like a chain. The ease of connection and access of IoT devices open doors for severe security issues especially with the large-scale distribution of heterogamous devices, their ability to connect to other devices without requesting permissions or even notifying their owners and probability of flooding these devices with severe security threats [12].

Handling security challenges in the IoT context should be a fundamental priority to increase adoption of IoT applications. Users need to be fully confident about the security of their IoT devices and related applications. They need to ensure that their devices are totally secured from various known threats as they become more integrated into people daily life's activities [13].

3.1 Security Requirements for IoT

Security of the IoT system can be assessed by employing classical security and risk analysis measures [14]. Typical CIA (Confidentiality, Integrity and Availability) security requirements should be employed in the IoT system.

Confidentiality means exchanging messages between a sender and receiver should be protected against any malicious or unauthenticated user [15]. For the IoT system, confidentiality need not only to be guaranteed inside the communication network but also when transmitting messages between various IoT devices. Integrity is used to guarantee the content of messages exchanged between the sender and receiver which is protected against any manipulation by an intruder without the receiver being able to track this manipulation. In the IoT system, the integrity check can be carried out at each node involved in the message exchange between the sender and receiver. Availability is used to guarantee that a malicious user is not capable of disrupting or harmfully affecting communication or quality of service provided by IoT devices or communication network [16].

Although CIA is essential for the IoT, there are other security requirements that are needed to be implemented for each level of the IoT architecture, as shown in Fig. 3. Node authentication is the main security issue for the physical layer to avoid unauthenticated node access and keep the communication channel between IoT nodes safe from any type of attack. Lightweight cryptographic algorithm and protocol is an important aspect to encrypt transmitted data especially for resources-constrained IoT devices [17].

For the connectivity or network layer, communication security measures are needed as well as identity authentication to prevent illegal nodes. Also, Distributed Denial of Service (DDoS) attack is common at this level, so there is a need to protect against DDOS attack in defenceless nodes in this layer, especially it is more severe in the IoT context [18]. For data abstraction, accumulation and edge-computing level,

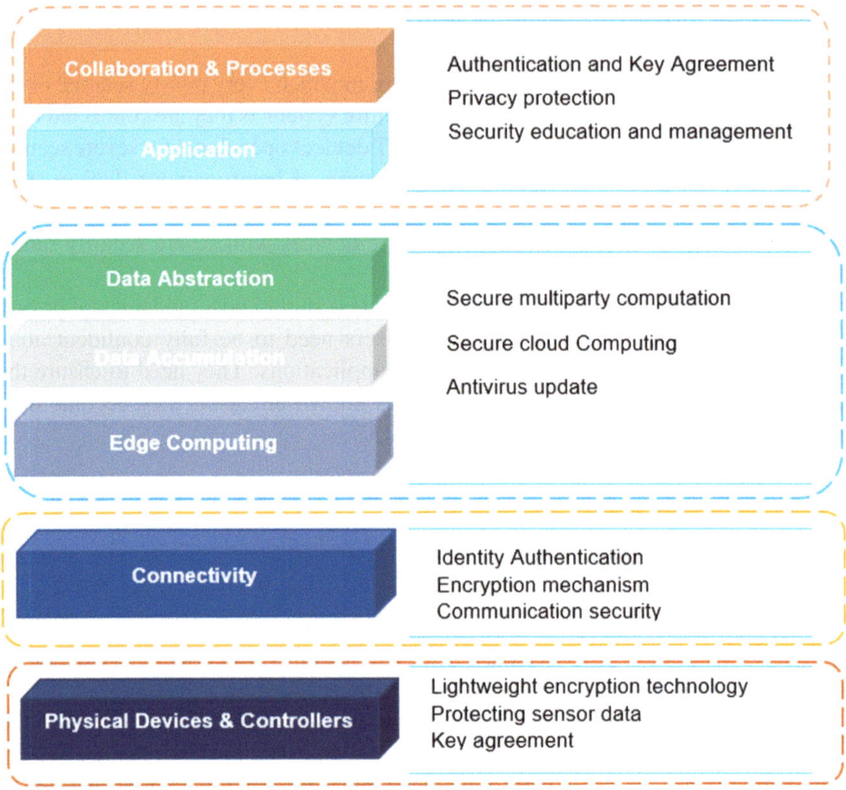

Fig. 3 Security requirements at each level of the IoT architecture

many application security mechanisms are needed to secure data stored in cloud computing. Strong encryption algorithms are needed besides an updated antivirus. While for the application and collaboration level, there is a need to adopt authentication and key agreement to protect user's privacy. Moreover, education and password management are essential for information security at this level [17].

3.2 Security by Design for IoT

Security by design is a novel approach suggested by several organizations to implement required security measures in the software and hardware development life cycle and not after detecting a security breach. The necessity to adopt security by design becomes essential to protect billions of IoT devices that are poorly secured against common security attacks. Since these devices are connected to the Internet, they become a weak point that can be exploited by any security attacker to steal sensitive information or disrupt the service. Also, the majority of these devices were built without security built into their system, making them easy targets for security attackers [1].

Security by design aims to protect the security of devices by the manufacturers. The user awareness of security has proved that it creates many vulnerabilities and threats that can affect people lives. Security by design can help the user to understand IoT security requirements and encourages them to make the right decisions to ensure their security and safety [19].

The UK government demand security by design in new products to address IoT security. The government argued that companies should integrate sufficient security mechanisms into their IoT devices to protect them from potential threats [20]. The government is also looking into providing incentives for the IoT industry to promote security by design for vendors and provide more information about built-in device security for consumers at purchase. Their strategy includes encouraging companies and developers to build safety features into their products from the beginning, to ensure connected devices are secure in both the design phase and throughout the life cycle of various products.

3.3 Best Practice for Securing IoT Devices

Security concerns associated with IoT devices create potential risks in our life. Before the IoT, a security breach can lead to losing your money, but with IoT, security attack can literally result in losing your life. Securing IoT devices requires taking a set of best practices that include the following:

- **Hardware Tamper Resistant**: Keeping IoT devices isolated and only certain people have physical access to it are the major steps to make your IoT devices tamper proof or tamper evident. Also, IoT device hardening with physical security such as blocking unused ports and covering camera are good points to prevent potential attackers from reaching your data [21].
- **Strong Authentication**: Many IoT users still use weak and default passwords without any update. Manufacturers should ask the customer to update the default one with strong passwords before using the device. In addition, alternative ways to recognize devices identity and trust are needed since username and password are not realistic for every device, especially for Machine-to-Machine (M2M) communication which starts to grow significantly [22].
- **Firmware Updates**: IoT devices must be patchable or upgradable with a proper digital signature. There are several serious threats on the Internet that affect IoT devices. Vendors and service providers should plan for future upgrades of devices' software to keep it up to date. These updates required to be accomplished on a time basis or subject to the importance of the update [23].
- **Device Identity Spoofing**: The sending and receiving nodes should be identified as legitimate devices. Therefore, it is significant to secure against IoT device identity

spoofing since setting and handling unique identities have been difficult for IoT devices due to their small and lightweight size.

- **Dynamic Testing**: It is critical for IoT devices to go through testing and create the least standard measures for security. To test the security of IoT devices, there are two types; static and dynamic. In contrast to static testing that is concerned with discovering threats in software, dynamic testing can explore threats and vulnerabilities in both hardware and software [21].
- **Failover Design**: IoT devices should operate appropriately in the case of losing or disrupting Internet connectivity. However, few IoT devices are built to work with such failure situations such as the Internet continuity or data disconnections. Failover design is essential for IoT devices that include user safety, such as door lock mechanisms, video monitoring, and environmental monitors and alarms. These devices should have additional features in the case of disconnected operations [24].

3.4 IoT Security Attacks

The IoT system with distributed and dynamic nature creates weak communication channels which are used by malicious objects to exploit and open new threats regarding tracking, monitoring and reporting of the users' actions. The increase of IoT devices in our community has presented a set of security attacks that need to be addressed. There are four main types of attacks in the IoT system, physical, software, network and encryption attacks. This section provides a brief discussion of each type of attack and its common examples within IoT systems. Various security attacks in the IoT system are summarized in Fig. 4.

3.4.1 Physical Attacks

These types of attacks are concerned with the hardware elements of the IoT system in which the attacker requires to be physically near to the IoT system to run the attack. These attacks are relatively difficult to achieve because they require expensive substances [25]. Physical attacks can have different forms which include the following:

- **Node Tampering**: This attack targets the sensor node by physically damaging it or even replaces the entire node or part of its hardware to gain the access to sensitive information [26].
- **RF Interference on RFIDs**: This attack targets the availability of the IoT sensors. The attacker uses Radio Frequency Identification (RFID) tag to direct noise signals using the Radio Frequency (RF) signals used by RFIDs for communication. These signals interfere with RFID signals which affect the quality of communication [27].

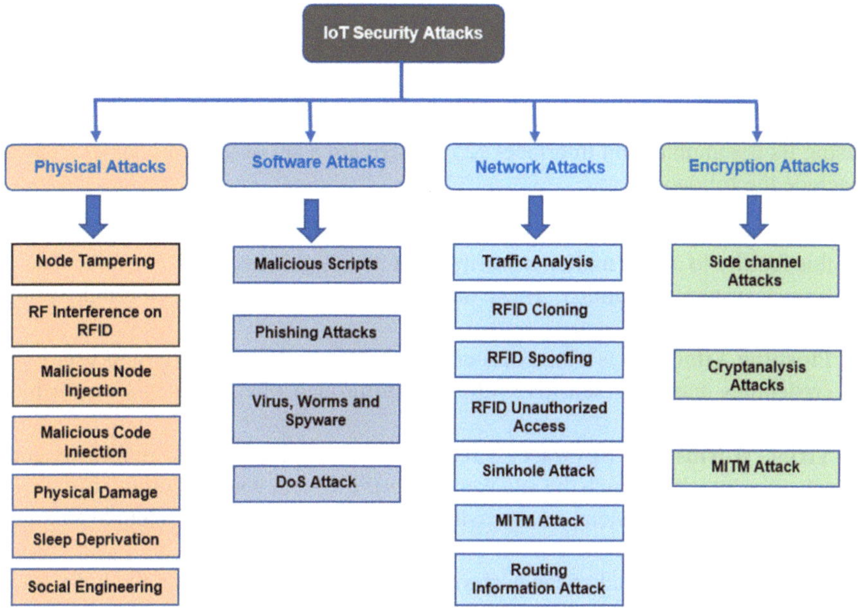

Fig. 4 Various security attacks in the IoT system

- **Malicious Node Injection**: The attacker gains the access to sensitive information by physically operating a new malicious node between communicating nodes of the IoT system, which allows the attacker to control all data flow between various nodes.
- **Malicious Code Injection**: This type of attack focuses on physically injecting the IoT node with malicious code that helps to gain access to the IoT system.
- **Physical Damage**: This type of attack is similar to node tampering in which the attacker physically damages IoT devices. This type of attack is difficult to achieve as it requires the attacker to reach area or building containing IoT devices to destroy it. The major difference between this attack and node tampering attack is that the attacker attempts to harm the IoT system directly to affect system availability and quality of service [26].
- **Sleep Deprivation**: Most sensors are operated through useable batteries that work according to sleep routine to enlarge their battery life. The sleep deprivation attack retains the nodes running at all times which leads to more energy feasting that results in shutting down of nodes after consuming battery energy [26].
- **Social Engineering**: The attacker uses the lack of security awareness of users to manipulate and gain access to the IoT system to collect sensitive information or to accomplish particular activities to serve his goals.

3.4.2 Software Attacks

Software attacks are the main cause of most security threats in almost all software systems. They target the weaknesses and threats found in the system implementation using its communication interfaces [25]. There are a set of software attacks which include the following:

- **Malicious Scripts**: Since the IoT system is linked to the Internet, the attacker uses this facility to create malicious scripts that aim to gain access to sensitive data or disturb system availability. These malicious scripts are executed through system users by wrong [28].
- **Phishing Attacks**: It is a kind of social engineering attack which targets user login credentials and other sensitive information through infected emails or phishing websites.
- **Virus, Worms and Spyware**: This type of attack is closed to malicious code injection attack in which the attacker injects the system with malicious software to gain access to the system, steal sensitive information or disrupt system availability [28].
- **DoS Attack**: An attacker can perform Denial of Service (DoS) on the IoT system across the application layer which affects all users of the IoT network. This type of attack also blocks legal users and gives the attacker the full access to sensitive data [25].

3.4.3 Network Attacks

The IoT system is a combination of networks interconnected together to transfer data between various IoT devices. Network attacks are concerned with the IoT network in which the attacker does not essentially require to be near to the network for the attack to operate. There are a set of network attacks which include the following:

- **Traffic Analysis Attacks**: This type of attack is concerned with sniffing out sensitive data and other types of data due to their wireless features. Moreover, in most attacks, it is necessary for the attacker to collect some network information before operating any attacks, and this is achieved by using a traffic analysis attack [29].
- **RFID Spoofing**: This type of attack is concerned with spoofing RFID signals to obtain data stored on an RFID tag. Then, the attacker uses the original tag ID to send his own data to appear to be from the original source, which enables the attacker to access the entire system as a legal node [30].
- **RFID Cloning**: This type of attack targets RFID tag by copying its own data to another RFID tag. Although the two RFID tags have identical data, it does not duplicate the original ID of the RFID [30].
- **RFID Unauthorized Access**: Due to the lack of appropriate authentication techniques in most RFID nodes, it is easy to be hacked by anyone allowing the intruder to read, edit or even delete data on RFID nodes.

- **Sinkhole Attack**: This type of attack targets the confidentiality of data and disrupt network service by discarding all packets instead of forwarding them to the desired destination [31].
- **MITM Attack**: Man-In-The-Middle (MITM) attack is operated by placing a malicious node between two communicating nodes which allow it to intercept and monitor all traffic sent between communicating nodes. Depending on the network communication protocols of the IoT system, the attacker does not need to be physically close to the network to run the attack [32].
- **Routing Information Attacks**: Routing table information is used by the network router to forward data to their desired destinations. Hence, this type of attack targets this table by spoofing or changing its contents which disrupt network service and most traffic will be discarded and error messages will be sent [30].

3.4.4 Encryption Attacks

The IoT system connects all objects through various communication channels. To protect the communication process, encryption algorithms are used. However, nothing is unbreakable. Encryption attacks are focused on breaching the encryption structure used in the IoT system [26]. There are a set of encryption attacks which include the following:

- **Side Channel Attacks**: This attack targets encryption devices in the IoT system using certain techniques to reach encryption and decryption keys used in the data encryption process.
- **Cryptanalysis Attacks**: If we suppose that the attacker already has the ciphertext or plaintext, then the attacker's goal becomes to find the encryption key by breaking the system encryption structure. There are several forms of cryptanalysis attacks such as chosen ciphertext, known-plaintext, ciphertext-only and chosen-plaintext attack [26].
- **MITM Attack**: For two nodes to communicate with each other on a secure communication channel using an encryption algorithm, they exchange encryption and decryption key. MITM attack tries to gain the access to this information by intercepting signals sent between two nodes and tries to execute a key exchange with each node separately, which enables the attacker to encrypt and decrypt any future signals between communicating nodes [32].

3.5 IoT Security Challenges

Like all new technologies, security issues are still the biggest problems that stand in the path of effective developments of the IoT system. There are several security challenges that need to be addressed to increase people trust in adopting IoT devices.

Fig. 5 Security challenges of the IoT system

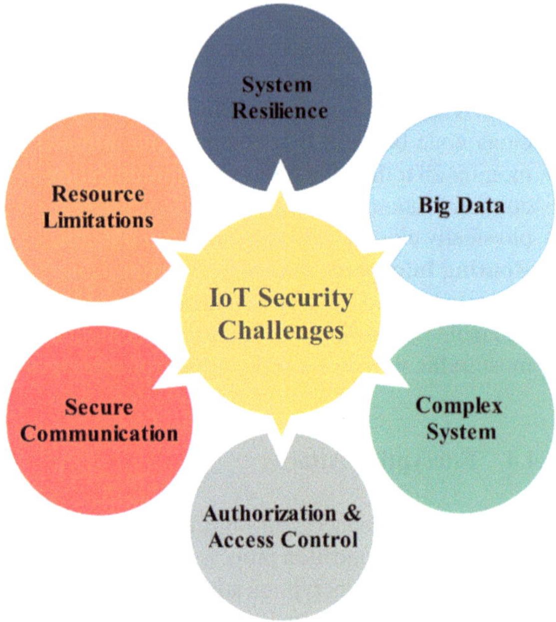

This section provides a brief discussion of common security challenges in the IoT system, as summarized in Fig. 5.

3.5.1 Resource Limitations

Most IoT devices have limited processing and storage capabilities, due to small and lightweight features which make them run on lower energy. Therefore, sophisticated security algorithms are not suitable for these constrained devices as they are not able to execute complex processing operations in real time. Instead, constrained devices typically only employ fast, lightweight encryption algorithms [33].

3.5.2 Big Data

As said earlier, the IoT system involves billions of devices which generate a huge amount of data. These data are variable in terms of structure and often arrive in real time. The volume, velocity and variety make storing and analysis process, which is used to generate meaningful information, a very complex task. The IoT is one of the main sources of big data. Using cloud computing can facilitate storing this huge amount of data for a long period of time. However, handling this massive data is a substantial challenge, as the entire performance of various applications is significantly dependent on the data management service. Moreover, one of the essential aspects of big data is data integrity. Ensuring the security of this huge amount of data is becoming difficult as data sources massively increased in a way that more security measures need to be adopted [34].

3.5.3 Authorization and Access Control

Providing an efficient authorization and access control mechanism for the IoT system is one of the major fundamentals to provide a secure system. IoT devices should gain access to services or applications only after providing their identities correctly. However, there are many problems associated with device authentication such as the use of weak or default passwords that lead to giving access to attackers who can manipulate device data or even physically damage it. Adopting security by design in IoT devices, enabling two-factor authentications and enforcing the use of strong passwords can help to address these challenges [35].

3.5.4 Secure Communication

Securing the IoT devices is not enough to ensure that the full security has been achieved in the IoT system. Instead, the communication channel connecting various communicating nodes such as IoT devices and cloud services needs to be protected from any attack. Most IoT devices send data in the plaintext format without encryption which make it an easy target to various types of network attacks. Hence, a proper encryption technique should be employed. Also, using separate networks can increase security through isolating devices and creating private communication channels.

3.5.5 System Resilience

Resilience is one of the main challenges that need to be addressed in the IoT system. System resilience refers to the ability of the system to respond to unpredicted attacks/situations without regressing. Hence, if some IoT devices are hacked, the system should be able to protect other network nodes from any attack.

3.5.6 Complex System

The IoT system involves billions of heterogeneous devices which make the management of this large-scale network a very difficult task to accomplish, especially with constraints associated with memory, energy and time. The more devices, people, interactions and interfaces, the more the risk of security breaches. In other words, with more variety and diversity in the IoT system, the challenges of managing all points in the network to maximize security become a difficult operation to achieve [2].

4 IoT Privacy

The IoT growth continues to add billions of new sensors and devices to the Internet, generating an enormous amount of information about people, including their locations, connections, shopping records, financial transactions, pictures, voices, conversations, health state, etc., with or without their consent. This huge amount of information makes retaining our privacy a difficult task [36].

The privacy in the IoT system can take many forms, but first, we need to define what privacy means. According to Westin [37], the privacy is defined as '*The claim of individuals, groups, or institutions to determine for themselves when, how and to what extent information about them is communicated to others*'.

Privacy is a notion that is associated with four main elements: information, communications, body and territory. Information privacy is related to various types of personal data collected and processed by an organization, such as financial and medical information, while the privacy of communication is concerned with protecting data sent between two communicating nodes using any communication medium. Body privacy is concerned with people's physical safety alongside any outside damage, whereas territorial privacy is concerned with building limits on physical space such as home, workplace and public places [38].

In the IoT context, protecting people privacy has become a very difficult task to achieve. This is because the data collection process is more passive, pervasive and less intrusive, which leads to making users less aware of being tracked. The potential risk of losing control over personal information is defined as a privacy threat. This threat is usually one of the key concerns of users and has an important effect on the adoption level of any new technology [39].

4.1 IoT Privacy Threats

One of the important characteristics of the IoT is the capability of objects to perceive and sense their environment. But this capability leads to tracking and monitoring user actions and activities which violate user privacy and results in many problems that can literally lead to losing people lives. This section provides a discussion of common privacy threats in the IoT system.

4.1.1 Identification

The IoT system is pervasive in nature that allows devices to sense and collect various types of data about users and their interactions with the environment. Typically, these data are processed at service providers, which are located outside of users' control.

Identification is the threat of relating an identifier (e.g., name, address) with private data about an individual. In the IoT, new technologies and interconnection of various techniques expand the threat of identification [40]. The use of a surveillance camera, in non-security contexts, is an example of such techniques, where customers' behaviour is studied for analysis and marketing. To address this issue, attribute-based authentication is recommended to minimize the data a device can collect in the IoT and maintain control over the disclosure of data.

4.1.2 Localization and Tracking

Localization and tracking are the threats of specifying and recording a person's location through time and space by different means such as cell phone location, Internet traffic or GPS data [40]. The availability of massive and complete spatial and spatiotemporal data has led to an increasing interest in using geographic data and incorporating spatial information analysis.

With the progress of the IoT system, several influences would apparently amplify the localization threats such as the expansion of location-aware applications and improvement of their accuracy, the ubiquity of data collection technology and interaction with IoT devices that record the identity, location and activity of the user.

4.1.3 Profiling

Profiling is the process of collecting and processing data about individuals' activities and actions over long periods to classify them according to some feature. The information is usually collected without permission from users and integrated with other personal data to create a more complete profile. Profiling is currently used in a large range of domains, for example, e-commerce, targeted advertising and credit scoring [41]. One of the risks associated with profiling is that personal information may be exposed to other users, as other users who share the same computer and browser may view one's targeted advertisement. Moreover, many users are disturbed by the mere awareness of being watched and tracked.

With the growth of the IoT, data collection incredibly increases quantitatively due to the explosion of data sources and connected devices. Furthermore, data will change also qualitatively as data is collected from previously inaccessible parts of people's private lives, for example, data collected by wearables and different devices at home [40].

4.1.4 Life-cycle Transitions

This kind of privacy threat refers to the disclosure of private information where the owner of a customer product is changed during its life cycle. Since consumer products that hold private information such as smartphones, cameras and laptops are mostly under the control of the same owner during their entire life cycle, this problem is not observed very often. However, as more and more everyday things will be connected and will contain private data, the risk for privacy disclosure due to change of owner will increase [42].

4.1.5 Inventory Attack

Inventory attacks are related to the illegitimate gathering of information about the existence and characteristics of things in a specific place. Inventory attacks can usually be performed by using the fingerprint of IoT devices, for instance, their communication speed, reaction time and so on. If the promise of the IoT will be fulfilled, all smart things will be addressable over the Internet, opening the opportunity for unauthorized entities to exploit this and create an inventory list of things belonging to a target. An inventory attack could be used for profiling individuals, since owning special items disclose private information about the owner [40].

4.1.6 Linkage

Linkage threat refers to uncontrolled disclosure of information due to combining separated data sources and linking different systems. Integrating various types of information about the individual reveals new facts which are not expected by the owner. The revealed information is considered a privacy breach [41].

Within the IoT context, the linkage threat will be increased due to integrating different organizations that establish a more heterogeneous and distributed system which will increase the system complexity and makes data collection operation less transparent [40].

4.2 Privacy-Preserving Solutions for IoT

Preserving the privacy of IoT devices should be one of the main priorities for the successful adoption and development of the IoT system. There are several approaches which have been suggested to preserve privacy. This section provides a brief discussion of these approaches to address the privacy issue in the IoT system.

- **Privacy by Design**: One valuable key to preserving privacy in the IoT environment is the privacy by design. The IoT customers should have the required features to control their own information and define who can access it. Currently, some companies use a sort of agreement that allows certain services to access data as desired. Therefore, built-in tools to preserve user's privacy are required to be built as an essential part of any product.
- **Privacy Awareness**: One of the main problems of privacy violation is the lack of public awareness. IoT users have to be fully aware of how to keep themselves protected against any types of privacy threats [43].

- **Data Minimization**: IoT service providers should employ the concept of data minimization by reducing personal data collection to only what is related to the service they introduce. They also need to retain the data only if they need it for the service [44].
- **Cryptographic Techniques**: One of the main solutions to preserve the privacy in IoT devices is employing the appropriate cryptographic technique to encrypt data. However, with limited storage and computation resources in IoT devices, this solution may be difficult to achieve [45].
- **Data Anonymization**: It is necessary after data collection that all unique identifiers such as social security number and driving license numbers should be removed from data records to remove the identity of the individuals in databases.
- **Access Control**: Providing an efficient access control model for the IoT system to enable smart things to provide fine-grained decisions is one of the solutions for preserving the privacy of IoT users.

5 IoT Safety

The IoT safety is one of the highest priorities to prevent the IoT system and its elements from producing physical damage or undesirable threat and protect the surrounding environment from such damage. In addition, safety-critical operations should be protected to maintain reliability in the IoT system. Ensuring safety and reliability in the IoT system is not an easy task. It requires not only building a consistent application but also developing new design approaches. Safety and security affect each other. Safety is concerned with physical damage of IoT devices and its surroundings. It is obvious that the physical system attached to a computer system generates a larger surface attack than a pure computer system. It also provides side channels attacks that enable intruders to detect and manipulate the computer system. Moreover, safety issues amplify the magnitudes of traditional security attacks [46].

Safety and security are integrated together at the design phase of the product life cycle and at runtime check for either physical system or computer system. Since IoT devices are connected to the Internet and new threats are exploited every day, a runtime check is more important to identify new threats and looking for the best method to mitigate against. Hence, the system needs to be monitored during operation to detect various threats [46].

The safety in the IoT system should be considered since a device may work safely in normal use, but if the device is hacked, the attacker will try to manipulate the functionality of the device causing harm to objects controlled by the device or compromise people approaching into contact with it [47].

There are serious safety issues coming with open and unused ports of IoT devices as it allows attackers to inject malicious codes causing damages to devices especially safety-critical devices. Therefore, this issue should be addressed in future product design to maintain physical security and safety of IoT devices.

6 IoT Ethics

Ethics is a branch of philosophy that defines human conduct and behaviour in the society. Ethics considers what is morally right or wrong, just or unjust, while rationally justifying our moral judgments. Ethics in the IoT context deal with defining the correct regulation for human activities towards others and themselves; hence, ethics can be considered as a way to define what is good and bad, right and wrong. With the IoT growth, it will possibly give rise to other moral dilemmas, especially as the technology continues to outperform the development of regulations and policies. The IoT will change everything about how society works and plays. Therefore, there is a need to develop an ethical framework that helps ensure the IoT is used for the good of humanity and not the other way around.

Due to the complexity, heterogeneity and large scale of the IoT system, new ideas and thoughts should be presented to define the appropriate regulation and policies for this complex environment. Ethical issues in the IoT are mainly caused by the expansion of IoT technologies [49]. In addition, as the community continues to explore the risks and opportunities associated with IoT-driven systems, attention to transparency and the ethics of these systems' use and behaviour needs to be a core part of the discussion. In addition, building ethical frameworks are needed to help to understand what is appropriate and inappropriate and what is good and bad. The mechanisms that enforce ethical IoT frameworks need to be relevant to an ecosystem that includes humans, autonomous and self-determining systems, devices, and virtual and physical environments [50].

A strong ethical standard will motivate companies to design smarter and more inclusively to avoid algorithmic issues and ensure global connectivity. When it comes down to it, every company is responsible for maintaining an ethical IoT foundation or consumers will deny access to their information resulting in a data deficit for companies. To get this right, leaders must consider the capacity of their IoT technology and how they can expedite access worldwide [48].

6.1 Ethical Design for IoT

With billions of IoT devices, the amount of data generated by these devices will be unpredictable. Integrating this amount of data with innovations and developments of efficient and effective big data analytics tools will change the people thinking about IoT and the huge economic progress that can be achieved using this data. On the other hand, there is still a lack of the appropriate ethics that regulate how these data can be collected without violating people's privacy. Therefore, an ethical design for future IoT devices and services is required to open various ethical options for users within the digital platform and make it act as an added value where user pay for it if he/she is willing to apply [48].

The ethical design in the IoT products is used as a means to authorize IoT consumers to manage and protect their personal data and other related information. In other words, IoT users will have the complete freedom to define their own ethical choices while interacting with IoT devices. All various ethical options and choices will be embedded in the algorithms that are created by programmers and developers. These choices will include different degrees of privacy and data protection to allow users to choose what is best for their purposes [50]. Since providing these new features are not free, an ethical IoT device will include additional cost to involve the implementation and deployment of ethical framing and ensure a higher level of freedom to IoT users. It will be available for users to decide to pay for these new ethical features or not [51].

According to W. Pollard [52], IoT devices involving ethical design should have the following features:

- The ability to manage and control the collection and distribution of personal data or services.
- The ability to apply different rules and policies regardless of time and space.
- The ability to support dynamic contexts such as home and office.
- The ability to observe, recognize and support relationships that need ethical options.

6.2 *Ethics Challenges in IoT*

Although the IoT system has been widely accepted in our society and billions of devices are existing, there are several issues to apply ethics in the IoT context. These challenges include the following:

- **Owner Identification**: The accurate identification of the owner of the data collected in a typical IoT system is difficult to define. Collecting various types of data without the user's consent or permission is a critical issue that needs to be addressed in the IoT system.
- **Public and Private Border Line**: The IoT system involves multiple sensors that collect both public and private data. In the absence of well-defined boundaries for users' information, the line between private and public information must be cleared and defined in various IoT applications.
- **People's Life Attacks**: In a pure computer system, the security breach can lead to data loss or physical damage to the computer system. While in the IoT system, as all our environment including our home, car, smart meter, etc. are connected within the IoT network, the IoT breach can literally affect people lives directly. For instance, an attacker can control home energy and cause serious damage to people living in that home [53].

7 Case Study: Smart Cities

The concept of a smart city is used to describe the better use of public resources to improve people quality of life using the unlimited benefits provided by the IoT system and at the same time decreasing operational costs of public administrations. The IoT provides numerous advantages in controlling and optimizing public services, such as lighting, maintenance of public areas, transport and parking, preservation of cultural heritage, surveillance and garbage collection. Moreover, with multiple sensors existing everywhere and different types of data collected from these devices, people awareness can be improved regarding the status of their city and encourage the active participation of the citizens in the management of public administration [54].

In this section, we provide the smart city as a case study to discuss different security threats and suggest novel solutions to mitigate against.

7.1 Security Threats in Smart Cities

Like all other IoT applications, smart cities provide an extensive range of vulnerabilities that can be exploited by attackers and other malicious actors causing serious damage to either people or physical devices. The security threats in the context of a smart city should not be ignored as it can affect productivity and efficiency of services provided by the smart city. There are several security threats in smart cities, some of the most common threats involve the following:

- **Data and Identity Theft**: Data created by unprotected smart city infrastructures such as parking garages and surveillance can be used to provide attackers with a huge amount of data to steal personal information that can be exploited for fake transactions and identity theft.
- **MITM Attack**: It is one of the common threats in smart cities in which an attacker injects a malicious node between two communicating nodes to steal conversation information. In smart cities, MITM attack on a smart valve can be used to intentionally cause wastewater overflow.
- **Device Hijacking**: In this type of attack, the attacker captures and controls a certain device without changing its basic functionality which makes it very difficult to be detected. In a smart city context, an attacker can exploit hijacked smart meters to launch ransomware on energy management systems [53].
- **Insecure Hardware**: Sensors are the starting point of any attack. If they are not tested appropriately, they will create major threats to the entire IoT system. The lack of hardware standardization of IoT devices creates several weak points that can be exploited by attackers.
- **Larger Attack Surface**: The large scale of a smart city network creates a large attack surface. Since smart cities contain thousands of systems and devices to control various services, any device in the smart city network is vulnerable and

can be attacked at any time. In addition, attacking a single device can possibly compromise the entire network [55].

- **Software Bugs**: Since smart cities contain thousands of systems and devices, a simple software bug can have an enormous effect on the system devices and applications.

7.2 Security Solutions for Smart City

Providing various security mechanisms to secure a smart city is a mandatory operation to keep the innovation of new services and applications that improve people lives and the quality of their lives. There are a set of security solutions for building a secure smart city. These solutions involve: mutual authentication, security monitoring and analysis, and data integrity and confidentiality. This section provides an overview of these security solution. In addition, Table 1 provides a summary of various security threats in different sectors of smart cities and suggested solutions.

- **Mutual Authentication**: Various types of devices connected to a smart city network should be authenticated before any data transmission occurs. This will validate the identity of communicating devices and ensure only legal devices are permitted to send and receive data. So, mutual authentication, where two entities device and service validate their identity to each other, can help to protect against malicious attacks [56].
- **Security Monitoring and Analysis**: The system data should be captured and monitored to detect potential security violations or potential security threats. Once a security threat is detected, appropriate actions according to system security policy should be performed [53].
- **Data Integrity and Confidentiality**: Smart cities use data to improve services and quality of life for citizens. This data should be reliable and accurate. In other words, integrity measures should be employed to ensure data is accurate and no manipulation occurs through the transmission process. Moreover, security measures should be employed to protect against the unauthorized disclosure of sensitive information.

8 Conclusion

The IoT has the capability to connect and communicate with almost all real-world objects over the Internet to increase information sharing. With the help of sensors, the IoT has the ability to collect, analyse and deploy a huge amount of data which in turn will be converted into meaningful information and knowledge that can be used to create new application and services to improve our quality of life. Security and privacy are considered to be the major issues in the IoT system. Providing a secure and

Table 1 Security threats and suggested solutions in smart cities, structured from some data provided in [55]

Sector	Security threats	Solutions
Smart building	• Systems failure • Controlling the fire system • Altering smart meters • Opening parking gates • Infection by malware • Damaging or controlling the lifts • Disabling water and electricity supplies	• Two-factor authentication • Threat and risk modelling • IoT forensics • Data backup and recovery solutions to guarantee reliability and continuity of services
Smart Transportation	• Sending wrong emergency messages • Stopping the vehicle's engine • Changing GPS signals • Disrupting the vehicle's emergency response system • Disrupting a vehicle's braking system	• Misbehaviour detection solutions • Pseudorandom identities • Public key infrastructure (PKI), digital certificates (ECDSA) • Data encryption solutions (ECIES and AES)
Healthcare	• Sending wrong information • Sending an emergency alert • Jamming attacks • Eavesdropping sensitive information • Disrupting the monitoring system • Disrupting the emergency services	• Secured Wi-Fi networks to guarantee safe handling of confidential information and personal data • Risk assessment
Energy	• Spoofing addresses and usernames • Unauthorized access and controls • Zero-day attacks • Denial of service and distributed denial of service (DDoS)	• Intrusion detection and prevention techniques • Risk assessment • Insider threat analysis • Cybercrime intelligence

privacy-preserving IoT system should be a compulsory task to continue its successful developments in our environment. In addition, safety plays an important role in the IoT system to provide a safe and reliable system and protect the IoT system and its components from causing an unacceptable risk or physical damage. In the same way, ethics and regulations when dealing with IoT data are needed to be defined since the technology continues to outperform the development of current regulations and policies. This chapter provided an overview of IoT security, privacy, safety and ethics. It discussed the architecture and essential characteristics of the IoT system. It also presented IoT security by highlighting security requirements, security by design, security attacks and security challenges. IoT privacy by investigating privacy threats and solutions to preserve privacy were also discussed. IoT safety and ethics were also discussed. In the end, a case study of the smart city was introduced to discuss security threats and suggested solutions to build a secure smart city.

References

1. Atlam, H.F., Walters, R.J., Wills, G.B.: Internet of things: state-of-the-art, challenges, applications, and open issues. Int. J. Intell. Comput. Res. **9**(3), 928–938 (2018)
2. Atlam, H.F., Walters, R.J., Wills, G.B.: Intelligence of Things: Opportunities & Challenges. 3rd Cloudification of the Internet of Things (CIoT), pp. 1–6 (2018)
3. Martin, P., Brohman, K.: CLOUDQUAL: a quality model for cloud services. IEEE Trans. Ind. Inf. **10**(2), 1527–1536 (2014)
4. Cerf, V., Ryan, P., Senges, M., Whitt, R.: IoT safety and security as shared responsibility. Bus. Inform. **1**, 7–19 (2016)
5. Shanbhag, R., Shankarmani, R.: Architecture for internet of things to minimize human intervention. In: 2015 International Conference on Advances in Computing, Communications and Informatics, ICACCI 2015, pp. 2348–2353 (2015)
6. Ashton, K.: That 'Internet of Things' Thing. RFID J., 4986 (2009)
7. ITU: The Internet of Things. ITU Internet Rep., p. 212 (2005)
8. ITU: Overview of the Internet of things. Ser. Y Glob. Inf. infrastructure, internet Protoc. Asp. next-generation networks - Fram. Funct. Archit. Model., p. 22 (2012)
9. Guillemin, P., Friess, P.: Internet of things strategic research roadmap. Eur. Comm. Inf. Soc. Media, Luxembourg (2009)
10. Stallings, W.: The internet of things: network and security architecture. Internet Protocol J. **18**(4), 2–24 (2015)
11. Cisco: The Internet of Things Reference Model. White Paper, pp. 1–12 (2014)
12. Atlam, H.F., Alenezi, A., Walters, R.J., Wills, G.B., Daniel, J.: Developing an adaptive Risk-based access control model for the Internet of Things. In: 2017 IEEE International Conference on Internet of Things (iThings) and IEEE Green Computing and Communications (GreenCom) and IEEE Cyber, Physical and Social Computing (CPSCom) and IEEE Smart Data (SmartData), no. June, pp. 655–661 (2017)
13. Iqbal, M.A., Olaleye, O.G., Bayoumi, M.A.: A review on internet of things (IoT): security and privacy requirements and the solution approaches. Global J. Comput. Sci. Technol.: E Network, Web & Secur. **16**(7) (2016)
14. Atlam, H.F., Alenezi, A., Hussein, R.K., Wills, G.B.: Validation of an adaptive risk-based access control model for the internet of things. Int. J. Comput. Network Inf. Secur., 26–35 (2018)
15. Maple, C.: Security and privacy in the internet of things. J. Cyber Policy **2**(2), 155–184 (2017)
16. Yu, Y., Kaiya, H., Yoshioka, N., Hu, Z., Washizaki, H., Xiong, Y., Hosseinian-Far, A.: Goal modelling for security problem matching and pattern enforcement. Int. J. Secure Softw. Eng. (IJSSE) **8**(3), 42–57 (2016)
17. Suo, H., Wan, J., Zou, C., Liu, J.: Security in the internet of things: a review. In: International Conference on Computer Science and Electronics Engineering (CCSEE 2012) vol. 3, pp. 648–651 (2012)
18. Abdur, M., Habib, S., Ali, M., Ullah, S.: Security issues in the internet of things (IoT): a comprehensive study. Int. J. Adv. Comput. Sci. Appl. **8**(6) (2017)
19. Theobald, M.: The Importance of Security by Design for IoT Devices (2018). https://www.redalertlabs.com/blog/the-importance-of-security-by-design-for-iot-devices. Accessed 20 Aug 2018
20. James, M.: Secure by Design: Improving the cybersecurity of consumer Internet of Things Report (2017)

21. George, C., Fink, G.A., Mandal, S., Hrivnak, C.: Internet of things (IoT) security best practices. IEEE Internet Technol. Policy Community White Paper, no. February (2017)
22. Atlam, H.F., Alenezi, A., Alharthi, A., Walters, R., Wills, G.B.: Integration of cloud computing with internet of things: challenges and open issues. In: 2017 IEEE International Conference on Internet of Things (iThings) and IEEE Green Computing and Communications (GreenCom) and IEEE Cyber, Physical and Social Computing (CPSCom) and IEEE Smart Data (SmartData), no. June, pp. 670–675 (2017)
23. Kvarda, L., Hnyk, P., Vojtech, L., Neruda, M.: Software implementation of secure firmware update in IoT concept. Adv. Electrical Electron. Eng. **15**(4), 626–632 (2017)
24. Venkatesh, J., Diego, S.: Scalable- application design for the IoT. IEEE Comput. Soc., 62–70 (2017)
25. Babar, S., Stango, A., Prasad, N., Sen, J., Prasad, R.: Proposed embedded security framework for Internet of Things (IoT). In: 2nd International Conference on Wireless Communication, Vehicular Technology, Information Theory and Aerospace & Electronic Systems Technology (Wireless VITAE), no. May 2014 (2011)
26. Sopori, D., Pawar, T., Patil, M., Ravindran, R.: Internet of things: security threats. Int. J. Adv. Res. Comput. Eng. Technol. (IJARCET) **6**(3), 263–267 (2017)
27. Deogirikar, J.: Security attacks in IoT : a Survey. In: International conference on I-SMAC (IoT in Social, Mobile, Analytics and Cloud), pp. 32–37 (2017)
28. Heer, T., Garcia-Morchon, O., Hummen, R., Keoh, S.L., Kumar, S.S., Wehrle, K.: Security challenges in the IP-based Internet of Things. Wireless Personal Commun. **61**(3), 527–542 (2011)
29. Khoo, B.: RFID As an enabler of the internet of things: issues of security and privacy. In: IEEE International Conferences on Internet of Things and Cyber, Physical and Social Computing (iThings/CPSCom 2011), pp. 709–712 (2011)
30. Mitrokotsa, A., Rieback, M.R., Tanenbaum, A.S.: Classifying RFID attacks and defenses. Inf. Syst. Front. **12**(5), 491–505 (2010)
31. Raju, I., Parwekar, P.: Detection of sinkhole attack in wireless sensor network. Adv. Intell. Syst. Comput. **381**(July), 629–636 (2016)
32. Padhy, R., Patra, M., Satapathy, S.: Cloud computing: security issues and research challenges. Int. J. Comput. Sci. Inf. Technol. Secur. (IJCSITS) **1**(2), 136–146 (2011)
33. Atlam, H.F., Attiya, G., El-Fishawy, N.: Integration of color and texture features in CBIR system. Int. J. Comput. Appl. **164**(3), 23–29 (2017)
34. Aman, W.: Modeling adaptive security in IoT Driven eHealth. In: Norwegian Information Security Conference (NISK 2013), pp. 61–69 (2013)
35. Atlam, H.F., Walters, R.J., Wills, G.B.: Fog computing and the internet of things: a review. Big Data Cognitive Comput. **2**(2), 1–18 (2018)
36. Atlam, H.F., Walters, R.J., Wills, G.B.: Internet of nano things : security issues and applications. In: 2018 2nd International Conference on Cloud and Big Data Computing, no. October, pp. 71–77 (2018)
37. Westin, A.F.: Privacy and Freedom. Atheneum, New York (1967)
38. Padilla-López, J.R., Chaaraoui, A.A., Flórez-Revuelta, F.: Visual privacy protection methods: A survey. Expert Syst. Appl. **42**(9), 4177–4195 (2015)
39. Atlam, H.F., Alenezi, A., Alassafi, M.O., Walters, R.J., Wills, G.B.: XACML for building access control policies in internet of things. In: Proceedings of the 3rd International Conference on Internet of Things, Big Data and Security (IoTBDS 2018), pp. 253–260. (2018)
40. Ziegeldorf, J.H., Morchon, O.G., Wehrle, K.: Privacy in the internet of things: Threats and challenges. Secur. Commun. Netwo. **7**(12), 2728–2742 (2014)
41. Toch, E., Wang, Y., Cranor, L.F.: Personalization and privacy: a survey of privacy risks and remedies in personalization-based systems. User Model. User-Adapted Interact. **22**(1–2), 203–220 (2012)
42. Aleisa, N., Renaud, K.: Privacy of the internet of things: a systematic literature review (Extended Discussion). ArXiv e-prints, pp. 1–10 (2016)

43. Atlam, H.F., Attiya, G., El-Fishawy, N.: Comparative study on CBIR based on color feature. Int. J. Comput. Appl. **78**(16), 975–8887 (2013)
44. Singh, J., Pasquier, T., Bacon, J., Ko, H., Eyers, D.: Twenty security considerations for cloud-supported internet of things. IEEE Internet Things J. **3**(3), 269–284 (2016)
45. Atlam, H.F., Alenezi, A., Walters, R., Wills, G.B.: An overview of risk estimation techniques in risk-based access control for the internet of things. In: Proceedings of the 2nd International Conference on Internet of Things, Big Data and Security (IoTBDS 2017), pp. 254–260 (2017)
46. Wolf, M., Serpanos, D.: Safety and security of cyber-physical and internet-of-things systems. Proc. IEEE **105**(6), 983–984 (2017)
47. Hussein, R.K., Alenezi, A., Atlam, H.F., Mohammed, M.Q., Walters, R.J., Wills, G.B.: Toward confirming a framework for securing the virtual machine image in cloud computing. Adv. Sci. Technol. Eng. Syst. **2**(4), 44–50 (2017)
48. Popescul, D., Georgescu, M.: Internet of things—some ethical issues. USV Ann. Econ. Public Adm. **13**(2), 208–214 (2013)
49. Alenezi, A., Zulkipli, N. H.N., Atlam, H.F., Walters, R.J., Wills, G.B.: The impact of cloud forensic readiness on security. In: 7th International Conference on Cloud Computing and Services Science, pp. 511–517 (2017)
50. Baldini, G., Botterman, M., Neisse, R., Tallacchini, M.: Ethical design in the internet of things. Sci. Eng. Ethics **24**(3), 905–925 (2018)
51. Atlam, H.F., Alenezi, A., Alassafi, M.O., Wills, G.B.: Blockchain with internet of things: benefits, challenges, and future directions. Int. J. Intell. Syst. Appl. June, pp. 40–48 (2018)
52. Pollard, W.: IoT governance, privacy and security issues. Eur. Res. Clust. Internet Things, 23–31 (2015)
53. Zanella, A., Bui, N., Castellani, A., Vangelista, L., Zorzi, M.: Internet of things for smart cities. IEEE Internet Things J. **1**(1), 22–32 (2014)
54. Ijaz, S., Ali, M., Khan, A., Ahmed, M.: Smart cities: a survey on security concerns. Int. J. Adv. Comput. Sci. Appl. **7**(2) (2016)
55. Kitchin, R., Dodge, M.: The (In)Security of smart cities: vulnerabilities, risks, mitigation, and prevention. J. Urban Technol., 1–19 (2017)
56. Khatoun, R., Zeadally, S.: Cybersecurity and privacy solutions in smart cities. IEEE Commun. Mag. **55**(3), 51–59 (2017)

CoAP—Application Layer Connection-Less Lightweight Protocol for the Internet of Things (IoT) and CoAP-IPSEC Security with DTLS Supporting CoAP

Aniruddha Bhattacharjya, Xiaofeng Zhong, Jing Wang and Xing Li

Abstract The Constrained Application Protocol (CoAP) is a standard web transfer protocol. The CoAP runs over UDP, resulting in an unreliable message transport. CoAP offers a request/response communication model among application endpoints. The Internet Protocol Security (IPsec) can offer various security services like limited traffic flow confidentiality, anti-replay mechanism, access control, confidentiality, connection-less integrity, and data origin authentication. One way to use IPSec to secure the CoAP transactions can be Encapsulating Security Payload Protocol [RFC 2406] (IPSec-ESP). It can be a special case, if the hardware provisions encryption at layer 2 (it is the situation with some IEEE 802.15.4 radio chips). Another way can be, the 6LowPAN (IPv6 over Low-power Wireless Personal Area Networks) extension, for using the IPSec with Authentication Header (AH) [RFC 2402] and Encapsulation Security Payload (ESP). To give more security to the major User Datagram Protocol (UDP) well-known applications, Datagram Transport Layer Security (DTLS) runs on top of UDP instead of Transmission Control Protocol (TCP). The DTLS offers automatic key management, confidentiality, authentication, and data integrity. It also provisions wide range of dissimilar cryptographic algorithms. We have found that providing end-to-end security is not so easy, so we have developed a Secure Hybrid RSA (SHRSA) cipher. At present, we are using it in personal messaging scheme, and it is able to provide end-to-end security with efficiency and lightweight features. Later, this cipher can be used in lightweight and efficient communication scenario of Internet of Things (IoT) and Internet of Everything (IoE).

Jing Wang died before publication of this work was completed.

A. Bhattacharjya (✉) · X. Zhong · J. Wang · X. Li
Department of Electronic Engineering, Beijing National Research Center for Information Science and Technology, Tsinghua University, Beijing, China
e-mail: li-an15@mails.tsinghua.edu.cn

X. Zhong
e-mail: zhongxf@tsinghua.edu.cn

J. Wang
e-mail: wangj@tsinghua.edu.cn

X. Li
e-mail: xing@cernet.edu.cn

M. Farsi et al. (eds.), *Digital Twin Technologies and Smart Cities*,
Internet of Things, https://doi.org/10.1007/978-3-030-18732-3_9

Keywords CoAP · Secure CoAP · DTLS · IPSec · Secure hybrid RSA (SHRSA) · SHRSA encryption · SHRSA decryption

1 Introduction

An unconstrained network (UCN) is classically signified by the Internet, while the Internet of Things (IoT) [1–20] comprising of a Low-Power Wireless Personal Area Network (LoWPAN) [6] signifies the constrained domain. An IoT gateway placed on the edge among the Constrained Network (CN) [21] and UCN [21] adapts the communication among these two domains. Its role typically encompasses the adaptation between dissimilar protocol layer implementations. Also called a border router, it carries out protocol translations vis-a-vis end-to-end loT security [1–4, 16–20, 22–38]. The gateway is usually an unconstrained device, which can be used for scaling down the functionalities from the UCN to CN domain. The gateway can be used for handling security settings in peripheral constrained networks.

To uphold the end-to-end method, the gateway necessitates to be invisible to the communicating endpoints. A node on the UCN can be either Hypertext Transfer Protocol (HTTP) enabled or only Constrained Application Protocol (CoAP) enabled. The communication protocols existing or being designed at the Institute of Electrical and Electronics Engineers (IEEE) and Internet Engineering Task Force (IETF) now empower a standardized protocol stack. The mechanisms founding this stack must thus empower Internet communications encompassing constrained sensing devices, whereas fulfilling the necessities of low-energy communications environments and the aims and the lifetime of IoT applications [22–27, 39–41]. In order to talk this issue for the IoT, the IETF has started the Constrained RESTful Environments (CoRE) working group, which aims at standardizing the incorporation of constrained devices with the Internet at service level. The CoRE proposal aims to permit the integration of constrained devices with the Internet, at service level. CoRE proposes the use of CoAP in constrained devices, a specialized RESTful Web transfer protocol.

1.1 Digital Twins

The idea of a digital avatar of a physical thing has now significant relevance in the recent years in view of IoT systems. The significant and relative example of this trend can be found in the Gartner's report titled "Top 10 Strategic Trends for 2017" (October 2016), Digital Twins was Number 5 strategic trend for 2017 in this report. Many vendors of IoT platforms have put some form of a digital twin into practice. These are usually named as twins, shadows, device virtualization, etc. The term "Digital Twin" was named by Dr. Michael Grieves at the University of Michigan during 2001–2002. In his paper, he brought the concept of a "Digital Twin" just the same as a virtual version of what has been manufactured. Dr. Grieves defined Digital

Twin Prototype (DTP), Digital Twin Instance (DTI) and Digital Twin Aggregate (DTA). The digital twin capability has the following three tools:

1. conceptualization,
2. comparison, and
3. collaboration.

In recent era, there have been significant advances in the capabilities and technologies of both the data gathering of the physical product and the formation and depiction of the virtual product, the *Digital Twin*. But during data communication again the *security of end-to-end path* is not much. So, we need to implement *end-to-end secure communication protocols* for these technologies related to IoT and Internet of Everything (IoE).

In this chapter, Sect. 1 has introduced our chapter with highlight on CoAP and Digital Twins technology. Then, Sect. 2 with its subsections have described the CoAP, its message structure, CoAP structure model, the CoAP-IPSEC (Internet Protocol Security (IPsec)) security, IPSEC issues, IPSEC and Datagram Transport Layer Security (DTLS) comparisons, and secure CoAP. After that, Sect. 3 has described DTLS supporting CoAP, some attacks, CoAP without DTLS and CoAP with DTLS, and some security issues. Our approach Secure Hybrid RSA (SHRSA) cipher is described in Sect. 4. Then, we have concluded the chapter in Sect. 5.

2 Constrained Application Protocol (CoAP)

CoAP is a specialized web transfer protocol aimed to be used by constrained devices in loT machine-to-machine (M2M) [14] applications. It is responsible for a client/server interaction model between application end points and comprises the same key functionalities of HTTP. So, CoAP can be easily interfaced with HTTP, resulting the web integration more simplified. Also, it is able to fulfill M2M critical necessities, for example, built-in discovery, simplicity, multicast support, and low overhead. loT nodes frequently have constraints concerning their resources, for example, computational power, memory size, and power management. Network communication, particularly wireless, also enforces additional limitations such as low bitrates, variable delays, and high packet losses. Due to the reason, the frames at the link layer are much smaller than the IPv6 MTU of 1280 bytes, extra adaptation techniques alike 6LoWPAN [6] for IEEE 802.15.4 [17] networks are essential, which further limits the network capacity. Yet, application layer protocols recurrently delegate security techniques to the transport layer, which benefits in attaining end-to-end security. The overhead caused by this security mechanism is very significant to the overall system performance. One such protocol is Datagram Transport Layer Security (DTLS), which furthermore has inbuilt binding within CoAP. Security is fundamental for the application areas. We should take care of the basic security services, for example, confidentiality, authentication, and freshness of secret keys between two communicating entities are there. Information exchanged in the network requisite to be protected end to end (E2E). To cope with these security necessities, CoAP offers DTLS and when DTLS NoSec mode is selected, the CoAP communication could be secured using IPSec at the network layer in a Low-power and Lossy Network (LLN). Nevertheless, DTLS was not intended for lossy networks and constrained devices, it has appeared as a vital candidate to deliver security in loT. Nevertheless, it cannot be employed as it is, ever since it is well thought out to be too heavy for use in constrained environments and networks such as loT. Thus, we have emerged numerous lightweight implementations of DTLS for use in loT nowadays.

Lightweight DTLS Implementation could be depending on employing any of the following techniques:

- preshared key (PSK),
- raw public key, and
- certificates.

The CoAP protocol defines bindings to DTLS to secure CoAP messages, together with few mandatory minimal configurations suitable for constrained environments. The acceptance of DTLS implies that security is reinforced at the transport layer, rather than being designed in the context of the application layer protocol. DTLS provides promises in terms of confidentiality, integrity, authentication, and non-repudiation for application layer communications using CoAP. CoAP is an application layer protocol. Also, it is a connection-less lightweight protocol for the IoT [1–20]. Some of the examples can be smart energy and building automation. The CoAP runs over UDP, resulting in non-reliable message transport. Another highlighting point is that it is not session based, and along with that the CoAP can tackle loss or delayed delivery of messages. CoAP offers a request/response communication model among application end points. It also has built-in discovery of services and resources support. The CoAP comprises significant conceptions of the web, such as extensible header options, Uniform Resource Identifier (URI), RESTful interaction, etc.

2.1 CoAP Massage Structure

CoAP's special ability is that it can effortlessly interface with HTTP for incorporation with the web, and at the same time, meeting specialized necessities required for constrained environments like very low overhead and multicast support. CoAP message structure [RFC 768] is shown in Fig. 1.

The first byte encompasses the protocol version Ver, a type field T (token), and TKL (token length). The T is a type field consisting of basic message type information. TKL represents the size in bytes of the Token field. Then, we have the Code field. The Code field encompasses more specific message type information. Then, we have Message ID field. The Message ID field is a unique ID. The work of this unique ID is to track messages and distinguish likely duplications . To match request and

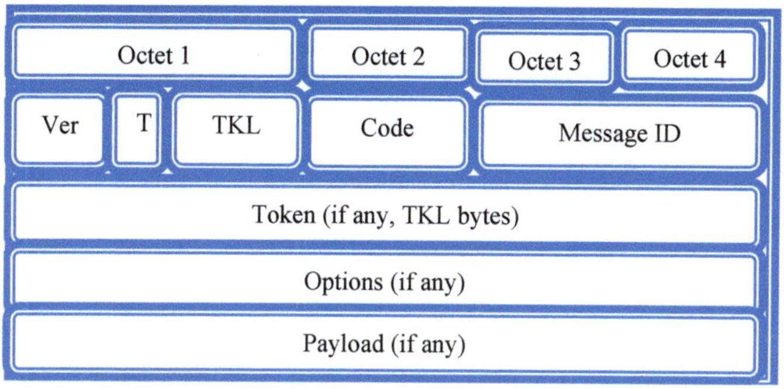

Fig. 1 CoAP message format

response messages, the optional Token field can be used. The value of this Token must be produced at random, and in addition to that it should be unique for each request. The field varieties are in between 0 and 8 bytes in size. These varieties of field are actually for making CoAP more robust to battle the IP-spoofing attacks. We should use this just in case, security is not offered at the transport layer. Moreover, more than a few dissimilar CoAP options have been well defined. Now, it is possible to state a list of them in line with a Type-Length-Content scheme. At last part of the structure of CoAP message, it has the Payload field. As we know, the IETF CoRE working group has projected the CoAP as a new application-level protocol for constrained devices. But astonishingly, the CoAP has no security measures, but nowadays, research works have projected positioning the DTLS or Internet Protocol Security (IPsec) protocols to offer a secure CoAP.

2.2 *CoAP Structure Model*

The CoAP is ideally made for constrained networks, such as 6LowPAN networks. The CoAP has the ability to use same methods as HTTP use, i.e., GET, POST, PUT, so it is well understood that the CoAP can take all the advantages from existing web-based technologies. We all know that the CoAP use User Datagram Protocol [RFC 768] (UDP) as a transport layer protocol instead of TCP. But this provides a very good advantage. The advantage is that for the reason of connection-less nature of the UDP, the CoAP can offer a lightweight reliability mechanism. It can be possible by dividing CoAP protocol into two layers as depicted in Fig. 2.

As shown in Fig. 2, the request/response layer is accountable for altering the resources by outlining methods (i.e., GET, PUT, DELETE, and POST). The Transaction layer recognizes the reliability technique, when processing messages, and along with that offer messages duplication detection. In the Transaction layer, we can have four types of messages, which are given below:

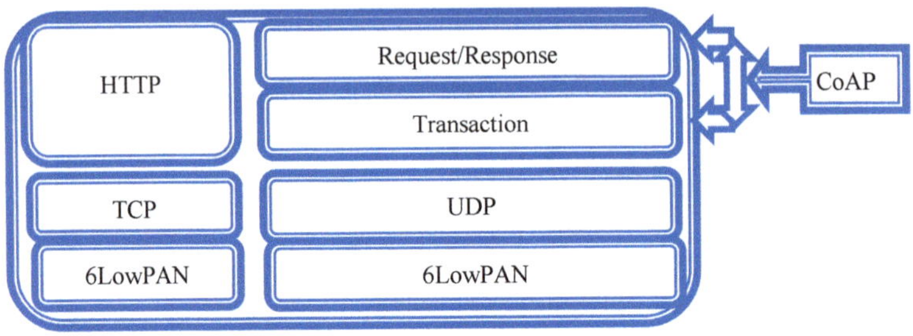

Fig. 2 CoAP and HTTP protocols stack

- Acknowledgment (to ACK. CON. Messages),
- reset (message is received but could not be processed),
- confirmable (Acknowledgment is required), and
- non-confirmable (no ACK is required).

The CoAP has many more abilities, which are essential features for IoTs environs [1–20], like URI and content-type provisions, low header overhead, simple parsing method, multicast support, and asynchronous message exchanges. We know that the CoAP's message could be CON (confirmable), NON (non-confirmable), ACK (Acknowledgement), and RST (Reset). A token is in use in the CoAP [RFC 768] for finding match for each response to its corresponding request. Messages are always exchanged in an asynchronous way. Messages always carry the semantics, responses, and requests.

The reliability is integrated into CoAP, and due to the reason that CoAP is bound to UDP. This reliability makes the CoAP more convenient and it is lightweight also. Reliability mechanism of the CoAP has the following features:

I. Very simple Stop-and-Wait Retransmission with exponential back off, for CON messages.
II. Detection of duplication for CON and Non-CON messages.

Therefore, it is well understood that the messages will be exchanged either reliably or non-reliably, liable to the option indicated in the GET request header. Figure 3 has shown that, if the resource is accessible in the time of dealing out the CON-Request, then the server will send the reply in a piggybacked manner with an ACK message.

Another case can be, if the server is not able to reply instantly to CON-Request message, may be for the reason of the lack of proper response, then the server in a simple way recognizes the request with an empty ACK message. In a situation, when the resources are available, then the server will send the reply in a new CON message. In turn of this CON message, the client will acknowledge. Figure 4 has shown this process.

The CoAP Internet draft has projected that two security protocols, DTLS and IPsec, can be in use to secure the CoAP network along with its traffic. But in later stage, the Internet Protocol Security (IPsec) protocol has been taken out from CoAP

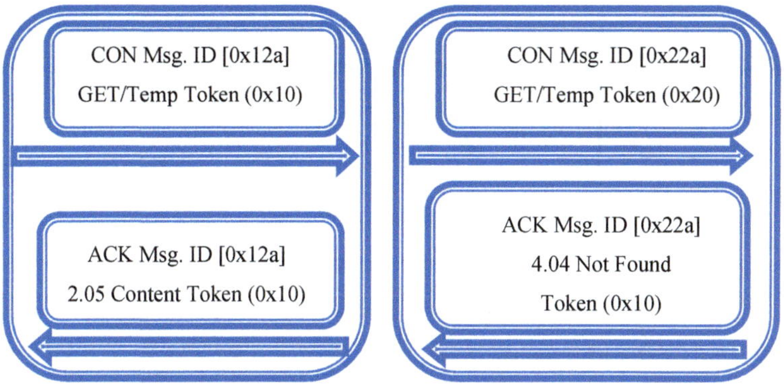

Fig. 3 Two GET requests with piggybacked responses

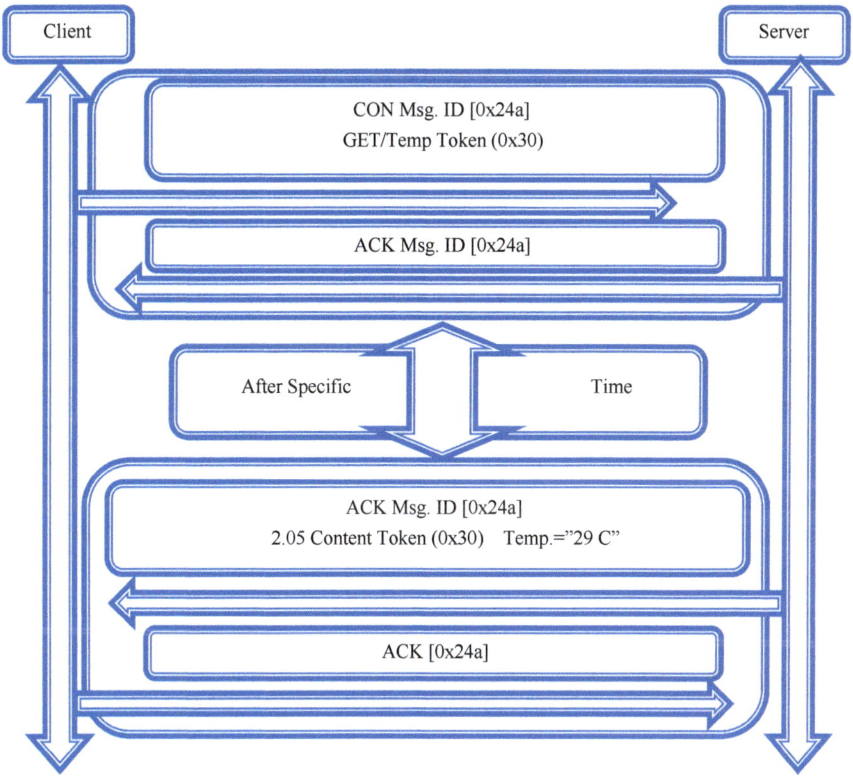

Fig. 4 GET requests with separate responses

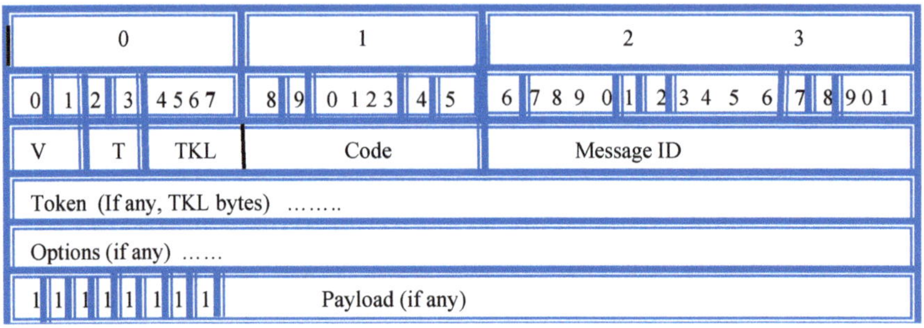

Fig. 5 Original CoAP packet format

draft and made an Internet draft. Moreover, few proposals have been published like either DTLS based or IPSec based with concern to CoAP security.

One of the good solutions to get more reliable access control framework for IoT can be merging other access control systems, like Kerberos and RADIUS with the CoAP protocol, to get a reliable access control framework for IoT.

Some special highlights of the CoAP are as follows:

- The URI format permits the use of standard and specialized service end points. One example can be, like the resource discovery, which is well defined in RFC 5785 that uses the well-known*/core* path and the CoRE link format.
- CoAP also permits to send very big messages with a stop-and-wait mechanism. This special mechanism is denoted as "blockwise transfers" (simply divide messages and transporting them with a reference order).

The CoAP packet format has a maximum length of 1400 bytes. Typically, the header is having a length of 32 bits (2 for the version control, 2 for message type, 4 for token length, 9 for the message code, and 16 for the message ID). The CoAP packet format is shown in Fig. 5.

In Fig. 6, we have shown that if we put more light on options format the format will be as shown below.

2.3 CoAP-IPSEC Security

We know that IPsec is a layer 3 protocol. It is ideal for use with IPv6, but in later stage, it is now can be used for IPv4. It can protect application and transport layers' applications, but good thing is that it is not an application-dependent protocol. The reason for this independence is that the IPSec is integrated into the kernel resulting in transparency to the applications. For the reason of this transparency, Transport Layer Security (TLS) and Secure/Multipurpose Internet Mail Extensions (S/MIME) [RFC 3851] can be used by IPSec. The IPSec can offer various security services like limited traffic flow confidentiality, anti-replay mechanism, access control, confidentiality,

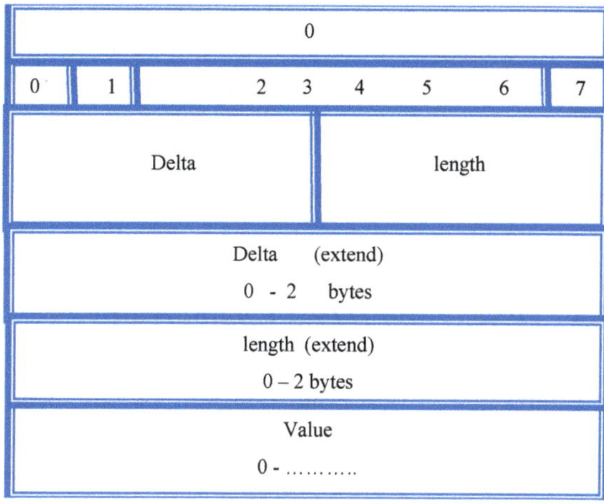

Fig. 6 CoAP option format

connection-less integrity, and data origin authentication. One way to use IPSec, to secure the CoAP transactions, can be Encapsulating Security Payload Protocol [RFC 2406] (IPSec-ESP). It can be a special case, if the hardware provisions encryption at layer 2 (it is the situation with some IEEE 802.15.4 radio chips). Another way can be the 6LowPAN extension, for using the IPSec with Authentication Header (AH) [RFC 2402] or Encapsulating Security Payload (ESP).

2.4 IPSEC Issues

There are some issues with IPsec.

First point is that basically the IPSec and DTLS were not considered for the constrained environs. At that time, the constraints were not considered in the IPSec/DTLS designs.

Second point is that IPSec has identified problems for making use of Network Address Translation (NAT) and/or Port Address Translation (PAT).

Third point is that performance of the network gets worse, when communicating small packets, as the encryption procedure of IPSec produces a large overhead.

Fourth point is that Security Associations (SA) has an issue, the mobility. The Security Parameter Index (SPI), Destination IP Address, and security protocol identifier identify the SA, uniquely. Now, in this case issue is that, if a node alters its IP address afterward the formation of the SA, then new SA prerequisites to be formed, which will give unnecessary performance degradation.

Fifth point is that IPSec is inserted in the IP stack, so any alterations will have need of kernel level.

Sixth point is that Configuring/Managing/Troubleshooting IPSec and Internet Key Exchange (IKE) are very composite tasks. It is well understood that enormous number of constrained devices are taking part in the network. Any wrong configuration security parameters of IPSec could give security holes or performance problems.

Seventh point is that every scenario/node cannot be supported by IPSec. Simply to understand, the support of IPSec for multicast communication is problematic.

Last but not the least, as per the CoAP's draft, it is promising to use IPSec (ESP) with layer 2 encryption hardware. It provisions the use of AES-CBC (128-bit keys).

2.5 IPSEC and DTLS Comparisons

A comparison of IPSec and DTLS in various security dimensions is described in Table 1.

Also, apart from the above issues, the DTLS and Internet Protocol Security (IPsec) are not the most enhanced resolutions to *offer proper protection to CoAP* for many reasons. The reasons are as follows:

1. IPSec and DTLS necessitate extra messages, to work for the security parameters and form the Security Associations (SAs). But the overhead and drain out of the resources of the constrained devices will be increased much more. This is very problematic for the mobile types of the devices in the IoTs.
2. The *Second point* is that if we think about the environs of the communication among two dissimilar networks, the ideal security resolution depends on either IPSec or Datagram Transport Layer Security (DTLS). This point toward the existence and provision of these protocols, in both the source and destination networks. But this ideal idea cannot be realistic in many circumstances, particularly when we think about the fact that the IPSec protocol has a compatibility problem with firewalls throughout the networks.

Table 1 A comparison of IPSec and DTLS in various security dimensions

Security dimension	IPSec	DTLS
Access control	No	No
Authentication	Yes	Partially—Server Only
Non-repudiation	Yes/No; as per the authentication method. PKI not supported by constrained devices	Yes/No; as per the authentication method. PKI not supported by constrained devices
Confidentiality	Yes	Yes
Communication Security	Yes	Yes
Integrity	Yes	Yes
Availability	Mitigation—No full defend	Yes—stateless cookie
Privacy	No	No

3. *Third issue* is that both IPSec and DTLS count on the Internet Key Exchange (IKE) and the Extensible Authentication Protocol (EAP) for setting up the secure association and sometimes any other. So, it is well understood that this point toward that all constrained devices vendors requisite to support these additional protocols (IKE and EAP).
4. *Fourth point* is that the IPSec and DTLS are aimed at securing connections among two static and remote devices. So, the IPSec and DTLS attempt to offer the most possible secure connection among the two ends, devoid of the QoS, the network trustworthiness, or any other restrictions on the end devices considerations. But in the environs of the constrained environment, there is a need for more dynamic and sensible actions that think about the constrained type of the end devices at the time of negotiating the security parameters.
5. The *fifth point* is that the IEEE 802.15.4 specification describes the payload should be 127 bytes as whole. Hence, if we use the DTLS as security protocol, to defend CoAP exchanges, 13 bytes (out of the 127 bytes of IEEE 802.15.4 frame) has to be assigned for DTLS record. Also, 25 bytes has to be used for link layer addressing information, 10 bytes for 6LowPAN addressing, and along with that the 4 bytes of CoAP header. So, as an outcome, only 75 bytes are available for application layer payload. But it is not sufficient space for communicating actual data. Subsequently, one big piece of data (bigger than 75 bytes) will use additional resources from the nodes and the network itself. The reason is that it will be broken into several pieces and sent twice. Hence, some header compression mechanisms are good solution, at the exact cases where needed. The compressing and decompressing necessities are the reason for more constraints to the nodes and network resources.
6. The *Sixth point* is that in the case of DTLS, some applications might necessitate security services, to be more and more customized in relation to the application or scenarios requirements. Nevertheless, if the security was applied as per the requirements of the application or scenario, it would offer to decrease the usage of existing resources and would increase the network enactment.
7. *Last but not the least*, in the Internet draft of "Datagram *Transport Layer Security in Constrained Environments*", the authors have pointed out seven prospective problems, correlated to DTLS protocol, if employed in constrained environs. The authors also have pointed out some projected workaround, to resolve these problems. Still, much works are required to make the DTLS perfect for making it a good and prospective security resolution for IoTs.

2.6 Secure CoAP

The *Secure CoAP (S-CoAP)* is a secure variant of CoAP. In S-CoAP, the security technique is actually an integrated part of the protocol itself. With S-CoAP, security measures will be integrated into the plain CoAP transactions. Therefore, one of the good features is that it will have its own compromise stage that thinks through the limits of the constrained devices. The S-CoAP prerequisites to offer security for normal connection setup, in addition to that, for the case of mobility also. In other words, the advantage is that the security will be integral part of the CoAP protocol. It is well understood that this security is offered by other standards, so the S-CoAP should be capable to function across numerous sites and networks.

3 Datagram Transport Layer Security (DTLS) Supporting CoAP

The DTLS protocol is UDP based. The DTLS is comprised of four protocols: *the Handshake protocol, Alert protocol, the Change Cipher Spec protocol,* and *the Record protocol.* The DTLS protocol offers message fragmentation at the handshake layer. This enables the DTLS to get rid of message fragmentation in the network layer. These fragmented packets bring many problems, like data loss rate increases and unnecessary delays made by packet retransmission. Therefore, the results indicate that LLN conditions are worse. The main burden to a memory-constrained device is to reunite a fragmented message packet, due to the reason that devices must retain fragmented pieces of the message in the buffer unless until all the pieces reach. To resolve these issues, the DTLS in Constrained Environments (DICE) standard WG was shaped. Nevertheless, definite solutions have not been projected yet. Therefore, it is a well-known thing that to decrease the load on memory of the devices used in making an IoT environs [1–20], lightweight DTLS was projected. *Lightweight DTLS* is able to decrease the DTLS code size for decreasing the burden on constrained memory of a device. Another way to reduce the load can be by decreasing the transmitted message size by compressing the DTLS header.

URI based on a CoAP communication environment having a RESTful structure is a good practical approach.

3.1 Some Attacks

Let us now discuss some of the issues about attacks for these systems.

1. ***Secure Service Manager (SSM) Spoofing Attack***

If an attacker is the SSM, then most dangerous thing is that the attacker can acquire all the information about the session, due to the reason of delegating the DTLS handshake. So, there is a chance that the encrypted data among end nodes can be exposed to the attacker. A good solution can be the use of PSK_DN (which is shared between the SSM and a constrained device in the bootstrapping phase). This is a perfect solution for protecting from SSM Spoofing Attack. The good reason for this protection of the SSM Spoofing Attack is that the data is encrypted by the use of

PSK_DN and then sent, and the attacker cannot deceive a constrained device and cannot get the right to use the encrypted data.

2. ***Semi end-to-end Security***

We have to ensure end-to-end security. The SSM can acquire all session information by just delegating the DTLS handshake. As we know the encrypted session information is sent to a constrained device instantly, but the SSM does not do the accumulation of session information. Therefore, it is well understood that end nodes joining in the DTLS communication will encrypt and decrypt data themselves only. The SSM is only responsible for the data relay after sending the session information to the constrained device. In this kind of system, the executor of the encryption and decryption is the end node, in the DTLS communication. There is one obligatory thing; the SSM must trust the preregistered device, for example, smartphone of user. So, as an outcome, we get an end-to-end security (semi E2E security exactly) definitely.

3. ***Denial of Service***

The devices setting up IoT have low CPU performance and a small amount of memory. Therefore, sending a DTLS handshake request message to these low-memory and low-performance devices can seem to be a DoS attack, even supposing the request is from a legitimate user. Another case is that if an attacker transmits a DTLS handshake message straight to a constrained device with conditions in the LLN, then as an outcome, the devices become more dangerous. Hence, the SSM benefits to resolve the DoS issue by delegating the DTLS handshake. The SSM stops constrained devices from receiving a lot of messages directly.

4. ***Single Point of Failure***

Numerous methodologies applying delegation can give a single point of failure (SPOF). It is one of the utmost predictable, but serious difficulties in security field. We know that the SSM has a significant role of delegating DTLS handshake in place of numerous CoAP sensors. So, it is well understood that if the SSM is negotiated or fails, then all the sensors under the SSM cannot create a secure session with client or server, which are outer of the LLN.

A well-defined trust manager can somehow protect such an SPOF issue. The trust manager has the option to choose alternative authentic device as a new SSM. Then, he can broadcast associated information to his sensors. Only thing is that the SSM should be resource-rich device in smart home or smart building (e.g., smart healthcare devices, etc.). Another way can be a virtually applied SSM in cloud system. It is harder to compromise a virtual SSM in cloud, as it is operated and supervised by security manager, compared to attacking a home device or smartphone, which is operated by its usual user. One highlighting point is that here a secure registration method between the SSM and IoT devices [1–20] controlled by the SSM is there. Moreover, another supposition is that the secret key, which is common for both SSM and its devices, cannot be compromised. Future research can be designing and implementing a concrete secure system, with additional mechanisms including key revocation, secure bootstrapping, trust management, and so on [22–27, 29–44].

5. ***Fragmentation Attacks***

A packet fragmentation mechanism is a good resolution for dissimilar MTU size between Internet and LLN. An IPv6 adaptation layer, 6LowPAN, has a provision with a method to fragment large IPv6 packets into small frame. Normally, sensing data and control data for actuators can be small. Though DTLS handshake message is bigger in size than the maximum frame of LLN in size, for instance, IEEE 802.15.4 (i.e., 127 bytes). Particularly, DTLS fragmentation is unavoidable at the fourth flight of DTLS handshake. The reason is that it encompasses comparatively large size of certificate of server and key exchange message. We can send 27 DTLS fragmented datagrams in case of using TLS_ECDHE_ECDSA_WITH_AES_128_CCM_8 with Raw Public Key Certificate. Significant transmission overhead is the outcome from these fragmented datagrams since the header is added to each of the frames. But some other critical issues are that, due to the deficiency in authentication mechanism at 6LowPAN layer, it gives chance for attackers to try buffer reservation attack, fragment duplication attack, and fragmentation attacks. An attacker eavesdrops and modifies a fragmented frame in the middle of the wireless multi-hop link to launch the fragment duplication attack. At the time of receiving, the Target node cannot identify the altered frame. So, as an outcome, the attacker's just a single forged frame can stop successful reassemble execution of the Target node. Additionally, the Target node requisites to abandon all frames in the buffer and waits for retransmission once more, as outcome, resulting in the Denial-of-Service (DoS) attack. We know that the first frame retains a memory space for reassembling the original packet and it is indicated in the header (i.e., datagram size field) at the target node. Furthermore, the buffer reservation attack exploits this fact. The attack can be very simple, like the attack can be done by sending a forged start frame encompassing large number in the datagram size field. A good option with a good efficiency can be a scheme, which uses the SSM to delegate the DTLS handshake phase. For the constrained network like Low-Power Lossy Networks (LLNs), network overhead and delay and loss problems, due to fragmented handshake message packets, are resolved by delegating the handshake. For the constrained device, the device need not retain the fragmented handshake packets in the buffer up to the receiving all of them. In addition, DTLS communication devoid of any source code for a DTLS handshake can be used by a constrained device. Here, the end-to-end security is definite, as data encryption and decryption are done in the end node. Moreover, its more important feature is that the system can tackle an SSM spoofing attack and DoS attacks on a constrained device. Another highlight is that the SSM and the constrained device are tangibly distinct, but can virtually be considered one system in a trusted relation with a shared key. This shared key is preshared. In other words, this kind of scheme can benefit to deploy constrained devices in a secure manner in constrained environments [21].

3.2 CoAP Without DTLS and CoAP with DTLS

The DTLS protocol is nothing but an improved type of the very popular TLS protocol [RFC 5246]. To give more security to the major UDP well-known applications, for instance, Voice over IP/Session Initiation Protocol (VoIP/SIP), DTLS runs on top of UDP instead of TCP. This is a key difference. The DTLS offers automatic key management, confidentiality, authentication, and data integrity. It also provisions wide range of dissimilar cryptographic algorithms. As per the CoAP's draft, CoAP describes four security modes with the intention of achieving the security services, which is obligatory. They are NoSec, PreSharedKey, RawPublicKey, and Certificate. In case of NoSec mode, the packets are transferred usually as UDP datagrams over IP. The CoAP scheme has indicated this as coap://. In case of all other three security modes, security is attained by DTLS and the scheme is indicated by coaps://. Figures 7 and 8 have depicted the message interchange for two cases, CoAP without DTLS and CoAP with DTLS:

Now let us discuss some issues of the *DTLS supporting the CoAP*. *At first,* multicast communications are not offered by DTLS protocol, but it is an essential part of CoAP protocol and main feature in IoTs. *Second thing* is that the DTLS handshake protocol is not protected at all, and at anytime it can be attacked by the exhaustion attack of the resources of battery-powered device, may be with the stateless cookie also. As an outcome, the nodes could not work properly in the network and make interruption to the whole communication. *Third*, bitmap window can defend the DTLS from replay attack, still nodes have to obtain the packets first, then process and occasionally even forward them also. This attack could make the network flooded. So, good resolution can be filtering proxy, for instance, 6LoWPAN [6] Border Router (6LBR). Moreover, one point in this resolution is that possibility of running this kind of filtering on a 6LBR cannot protect all situations. Furthermore, handling the replied packets is energy consuming. *Forth* issue is Handshake phase which is strongly defenceless, ever since no end host has been authentic to the other end host. *Fifth* issue is that DTLS security advantages do not match with the CoAP. For example, the loss of a message in-flight necessitates the re-communication of all messages in-flight. But, if all messages in-flight are communicated together in

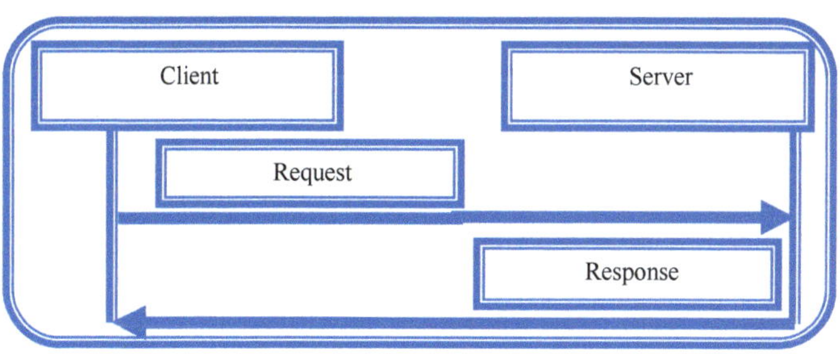

Fig. 7 CoAP request/response, one round trip without DTLS

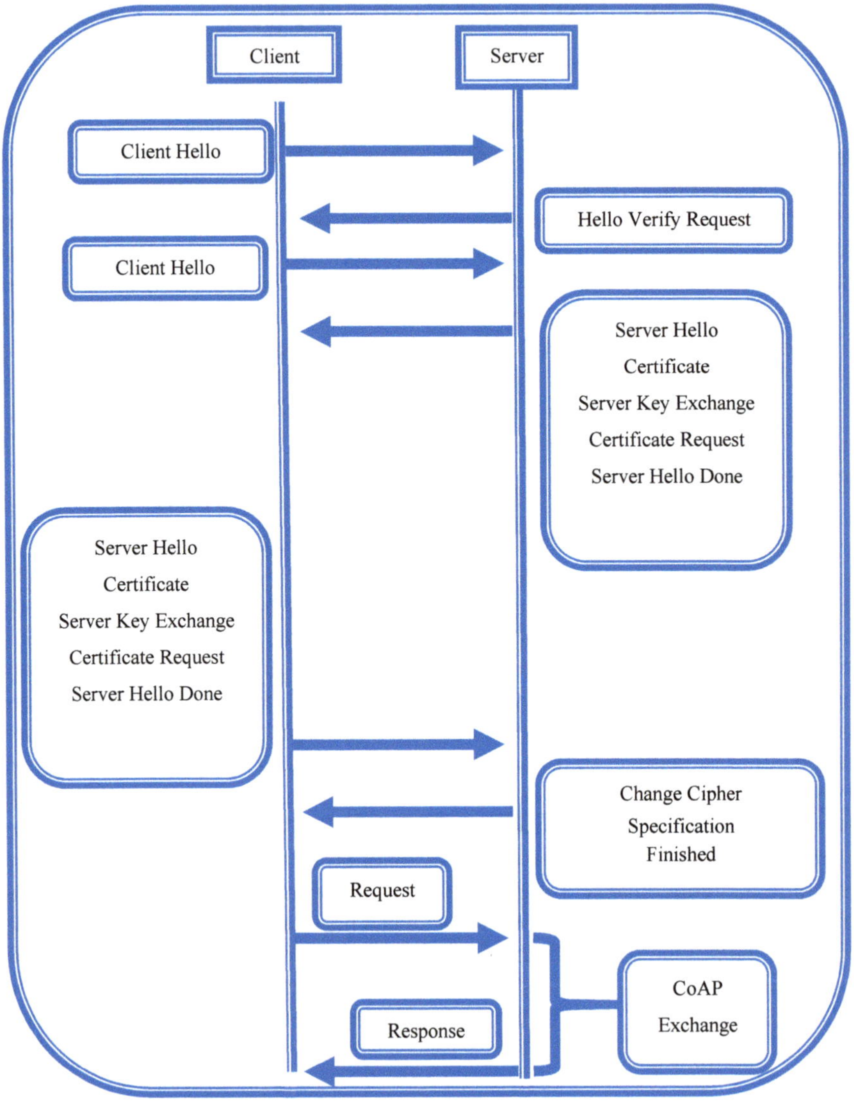

Fig. 8 CoAP request/response with DTLS, four round trips

a single UDP packet, it is good, but more resources are obligatory for dealing with large buffers. Additionally, if CoAP client prerequisites Internet access, which essentials the CoAP/HTTP mapping process, then, it is well understood that the DTLS handshake process will be big issue.

Mainly, it is not clear if a partial mapping among TLS and DTLS can be accomplished. This topic could also be more complex, since a CoAP client would not be capable to distinguish which device has started the request. *Last but not the least*, CoAP messages have two transactions (one round trip); one message starting at the client (request) and the other starting from the server (response). If DTLS is used in these two transactions processes, then we need four round trips, three round trips for DTLS (~40–50 Bytes), and additional one round trip for CoAP. It should be before CoAP's actual contents are exchanged.

3.3 Security Issues

Distributed IoT applications [22–27, 29–44] can use the CoAP at the application layer, with the intention of regaining the resources from sensing devices and in case of the autonomous communications, between WSN and Internet devices. CoAP can be used to empower the application layer RESTful communications with these sensing platforms. Therefore, this can be one of the foundations for the forthcoming great future of future IoT applications [22–27, 29–44]. The security in case of the CoAP has a major importance. The existing CoAP specification accepts Datagram Transport Layer Security (DTLS) at the transport layer security, for transparent secure CoAP communications at the application layer. DTLS offers end-to-end security. But, DTLS has conflicts with one functionality designed in CoAP, that is, the usage of proxies to help communications among the Internet and WSN communication domains. Another prospect for DTLS for CoAP necessitates the use of public key authentication by use of Elliptic Curve Cryptography (ECC) for the purpose of the authentication and key agreement.

The handshake is a big issue for the end-to-end security. The reason is that after completion of the authentication and key negotiation, the end-to-end security implementation issue can be resolved in the sensing device very efficiently with AES/CCM encryption. We know that the transparent interception and mediation of DTLS also give us advantages, other than permitting the ECC encryption to make provision for high security with CoAP. The end-to-end security's one of the key component can be the DTLS handshake. It permits mutual authentication and key agreement, within communicating both the parties. But it takes some more loads due to its high computation costs, so we should try to offload such costly computations. But when we are thinking this, we prerequisite to support sensing devices for moving freely in between several WSN domains. We have to take care about the matter that in the environs of a given IoT application, CoAP resources exist on sensing devices are securely reachable. The reachability with security should be regardless of the present location of the device. In parallel, there should not be any changes for CoAP and DTLS as maintained on such devices.

Even though IPsec can be used in the IoT [1–20], it is not principally intended for web protocols, for example, HTTP or CoAP. For web protocols one of the most common security solution is Transport Layer Security (TLS) or its forerunner Secure Sockets Layer (SSL). The connection-oriented TLS protocol can only be used over stream-oriented TCP that is not the favored method of communication for smart objects. By reason of lossy nature of low-power wireless networks, it is tough to keep a nonstop connection in 6LoWPAN networks. An adaptation of TLS for UDP known as Datagram TLS (DTLS) is in use nowadays. It initiates end-to-end security of dissimilar applications on one machine by functioning among the transport and application layers. DTLS offers replay protection, integrity, confidentiality, and authentication. It also offers defence against Denial-of-Service (DoS) attacks by using cookies. As we know DTLS offers application-level end-to-end security, it can only be in used over the UDP protocol and TLS is in use over TCP. The secure web protocol for the IoT, Secure CoAP dictates the use of DTLS as the underlaying security resolution for CoAP. It is well understood that we should permit DTLS support in the IoT [1–20]. Communication security defends the messages with confidentiality and integrity services; still many attacks are likely against networks mostly to breach availability security services. These attacks are intended to barge in networks by interrupting, for instance, the routing topology or by launching DoS attacks. Intrusion Detection Systems (IDS) are obligatory to identify impostors and malicious activities in the network, and firewalls are essential to stop unapproved right of entry to networks. In the IoT, 6LoWPAN networks are defenceless to a number of attacks from the Internet and from inner of the network. Also, 6LoWPAN networks itself can turn out to be source of attacks against Internet hosts, as it is reasonably easier to negotiate a resource-constrained wireless node than a typical Internet host. *As a result, we have found that it is not so easy to provide end-to-end security.*

4 Our Approach Secure Hybrid RSA (SHRSA) Cipher

Our approach is that we first took some of the RSA variants and insert them in our Secure Hybrid RSA (SHRSA) cipher with its own encryption and decryption [36–38, 45–48], with some other algorithms to resolve some of the major problems with main RSA, as shown in Fig. 9. As an outcome, we have developed an SHRSA messaging scheme with Secure Hybrid RSA end-to-end encryption and secure hybrid RSA decryption [36–38, 45–48].

Our scheme's encryption gives us following solutions:

1. Enable us to tackle dangerous known-plaintext attacks.
2. It is giving us stronger statistical complexity, meaning more security.

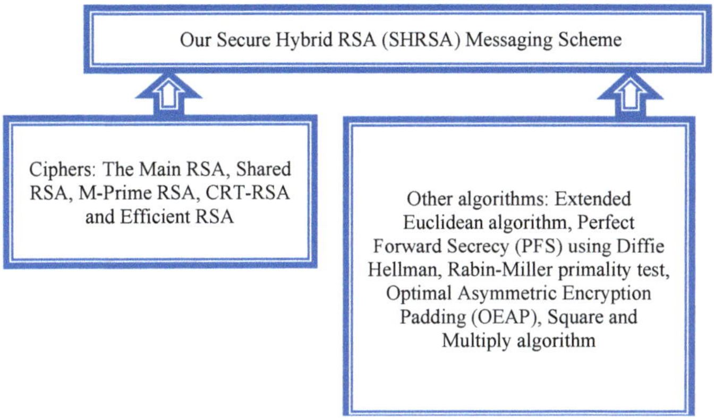

Fig. 9 Our SHRSA messaging scheme

3. It is giving us parallel protection to sniffing attacks and real-time key negotiation between each peer is given by Perfect Forward Secrecy (PFS) using Diffie–Hellman.
4. It is having more and more effortlessness for users with high speed by use of efficient RSA with extended Euclidean algorithm.
5. The computational rate due to encryption will not be different ominously from the original RSA. So, our hybrid RSA encryption is also having the encryption complexity $(3n_e\text{-}2)\ (n^2 + 2)$ alike main RSA.
6. It is enabling us to be protected from exploitation of multiplicative property and homomorphic property (meet-in-the-middle attack).
7. It is giving resolution for difficulty of the integer factorization problem of RSA and very computationally costly exponentiation modulo N problem.
8. 1024-bit key is giving us solution for the exploitation of certain key choices problems.
9. RSA's high computational cost is totally irrelevant in all aspects of the Internet of Everything (IoE) scenarios. This high cost owing to modular exponentiation is getting decreased by use of the square and multiply algorithm in our scheme's nine-layered protocol stacks.

Using Perfect Forward Secrecy (PFS) our SHRSA communication protocol is able to protect our messages from the following:

1. non-repudiation attack,
2. man-in-the-middle attack,
3. chosen cipher attack, and
4. replay attack.

OEAP insertion with random salts in our nine-layered SHRSA cipher is helping us to stop attacks like those given below:

- Algebraic attacks,
- The Hastad attack,
- Desmedt–Odlyzko attack,
- Related message attacks,
- Fixed pattern RSA signature forgery, and
- Two attacks by Bleichenbacher.

Our SHRSA decryption is not only giving solution to low modular complexity problem but also has made our SHRSA decryption stronger, more complex, more challenging, and difficult to be broken.

Our secure hybrid RSA messaging system [36–38, 45–48] is able to replace these following disadvantages of existing instant messaging schemes and protocols [36–38, 45–48]:

1. It is distributed no single point failure with SHRSA and it works peer to peer for all kind of users.
2. Secure hybrid RSA decryption is nine times faster than main RSA decryption.
3. Secure hybrid encryption and decryption are much more complex between each peer.
4. Pohlig–Hellman (PH) key exchange and Diffie–Hellman exchange key ensure three-way authentication peer to peer.
5. Optimal Asymmetric Encryption Padding (OEAP) with some random salts added on runtime with synchronize time gap protects chosen ciphertext attack and short plaintext attack, man-in-the-middle attack, and other attacks.
6. SHRSA works with any network with dual stack with native IPv4.
7. It has end-to-end encryption with full mesh topology.
8. No SSL is used and no external digital certificates are used, and we have our own hybrid RSA security, authentication, and integration with very strong confidentiality. It can replace all backlogs of SSL systems.
9. No default setting is shared with others.
10. No need of any third party.
11. It is more reliable, more efficient, and stronger due to variants of RSA integration.
12. It is protected to attacks like factorization of the RSA modulus n, message iteration attack, broadcast decryption by small exponent attack, broadcast decryption by common modulus attack, fault injection attack, the small difference between p and q attack and the finding eth root attack, mathematical attacks, and timing attacks. DoS and replay attacks are prevented also.
13. Brute-force attack is tackled by randomly changing the keys in synchronous time gap with 1024-bit value.
14. No need to install IMSecure.
15. No need of use of any password as we have our own three-way four-layered authentications for peers and then secure hybrid RSA encryption.
16. Our SHRSA messaging system works with an end-to-end encryption model with full mesh networked architecture to ensure pure peer-to-peer nature.

Also, this end-to-end secure SHRSA cipher can be used for securing the data, which are exchanged between real space and virtual space and information are processed in *Digital Twin concept model*, and those data and information are totally insecure presently.

Now, our SHRSA messaging system can be used in following real-time daily use:

1. In a distributed environment where multiple servers and multiple clients can communicate in a peer-to-peer manner with strong, reliable, and efficient end-to-end security.
2. Replacement of SSL/TLS was needed for personal messaging scenario, and it can be used (as SSL/TLS has several backlogs).
3. Our end-to-end user three ways authenticated encrypted messaging architecture based on SHRSA can be incorporated in the future Internet architectures like Choicenet, NEBULA, and eXpressive Internet Architecture (XIA) along with Sourceless Network Architecture and with binding of CoAP with DTLS.
4. Our architecture also affords a hassle-free, secure, peer-to-peer, unconventionally strong, and reliable platform with end-to-end encryption for people and organizations who are concerned about their privacy and security.
5. We have our own three-way four-layered authentications for peers and then SHRSA encryption, so in a scenario where we have the need to use any password (like AOL Instant Messenger (AIM), ICQ, MSN Messenger (Windows Messenger in XP), and Yahoo! Instant Messenger (YIM)) our SHRSA can be used without passwords.
6. In a scenario, where external digital certificates are used, our scheme can work without external digital certificates, as we have our own SHRSA security, authentication, and highly efficient architecture with very strong confidentiality.
7. In a scenario, where there is need of any third party (like Instant Messaging Key Exchange (IMKE) protocol), our SHRSA system can work well, as no need of third-party authentication, variants of RSA integrations give more reliable, more efficient, and stronger security.

5 Conclusions

In 1999, the Auto-ID Laboratory of Massachusetts Institute of Technology has introduced us thought of "the Internet of things". Then, in 2005, we had the "ITU Internet Reports: The Internet of Things". We need to develop the security structural design of the IoT, for the reason of offering information security defence for tag privacy, sensor data security, data transmission, etc. We need very deep systematic research on the transmission and information security of the core network depending on the IoT or networking industry security of the IoT. We have seen that recent works are simply adding safety methods in each layer. But this is not at all sufficient. We have already worked on forming lightweight hash functions, which depend on lightweight block ciphers. We know that AES-CCM (Advanced Encryption Standard (AES) CTR mode with CBC-MAC) and AES-GCM (Advanced Encryption Standard (AES) Galois/Counter Mode (GCM)) project's data integrity and confidentiality. Another way for optimization can be algorithm management in cross-layer architecture. Here, the reason for optimization is that numerous security mechanisms share one algorithm. The Internet Engineering Task Force intention is to execute Internet standards in the IoT. We have seen that many researchers have tweaked the IPsec protocol, for offering the network layer security between Internet hosts and constrained devices. But still some issues are hard to resolve. We all know that the IPsec prerequisites a shared password, for doing the encryption and decryption for appropriately all incoming and outgoing messages. But big issue is that if these passwords are static then it can be compromised after some thousand messages. For resolving this issue, the IKE and IKEv2 protocols were formed. These protocols promise a protected communication between two devices and are capable to generate new shared passwords, by use of circling derivative tactics. CoAP without DTLS and CoAP with DTLS can be an option. We use DTLS for protecting UDP packets (even over IPsec). By use of an initial handshake, it set the passwords. Then, the content of the UDP packet is encrypted (usually with TLS PSK over AES) and a header of 13 bytes is added. This process is done together with the Initialization Vectors (IV) (over 8 bytes for AES128), integrity values (8 bytes), and the padding prerequisite by the cipher suite. Still intensive researches are needed for perfect end-to-end security providing solutions in light of IoT. Our proposed SHRSA cipher can be used for efficient and lightweight solutions for end-to-end encrypted communication protocol use. It is unquestionable that the security of the Internet of Things is more than a technical problem, which also prerequisites a series of policies, laws and regulations, and perfect security management system for mutual collocation.

Acknowledgements This research is supported by National Natural Science Foundation of China (No. 61631013). We want to convey our gratitude and tribute to Late Prof. Wang Jing for his constant supervision and encouragement for this project.

References

1. Jara, A., Kafle, V., Skarmeta, A.: Secure and scalable mobility management scheme for the internet of things integration in the future internet architecture. Int. J. Ad Hoc Ubiquitous Comput. **13**(3–4), 228–242 (2013)
2. Li, S., Gong, P., Yang, Q., Li, M., Kong, J., Li, P.: A secure handshake scheme for mobile-hierarchy city intelligent transportation system. In: International Conference on Ubiquitous and Future Networks. ICUFN, Da Nang, pp. 190–191 (2013)
3. Kang, K.C., Pang, Z.B., Wang, C.C.: Security and privacy mechanism for health internet of things. J. China Univ. Posts Telecommun. **20**(Suppl 2), 64–68 (2013)
4. Goncalves, F., Macedo, J., Nicolau, M., Santos, A.: Security architecture for mobile e-health applications in medication control. In: 2013 21st International Conference on Software, Telecommunications and Computer Networks. SoftCOM, Primosten, pp. 1–8 (2013)
5. An, J., Gui, X., Zhang, W., Jiang, J., Yang, J.: Research on social relations cognitive model of mobile nodes in internet of things. J. Netw Comput Appl **36**(2), 799–810 (2013)

6. Kasinathan, P., Costamagna, G., Khaleel, H., Pastrone, C., Spirito, M.: Demo: an ids framework for internet of things empowered by 6lowpan, Berlin, Germany, pp. 1337–1339 (2013)
7. BETaaS Consortium (2014) BETaaS building the environment for the things as a service D2. 2. 2–Specification of the extended capabilities of the platform, pp. 1–61
8. IoT-A Consortium (2014) IoT-A unified requirements. http://www.iot-a.eu/public/requirements/. 31 Jan 2014
9. Gao, L., Bai, X.: A unified perspective on the factors influencing consumer acceptance of internet of things technology. Asia Pac. J. Mark. Logist. **26**(2), 211–231 1075 (2014)
10. Gazis, V.: Carlos Garcia Cordero, Emmanouil Vasilomanolakis, Panayotis Kikiras, and Alex Wiesmaier. Security perspectives for collaborative data acquisition in the internet of things. In: International Conference on Safety and Security in Internet of Things. Springer, New York 1079 (2014)
11. IoT-A Consortium (2014) IoT-A—Internet of things architecture. http://www.iot-a.eu/. 27 Jan 2014
12. Logvinov, O., Kraemer, B., Adams, C., Heiles, J., Stuebing G.: Mary Lynne Nielsen, and Brenda Mancuso. Standard for an architectural framework for the internet of things (IoT) IEEE P2413 Webinar Panelists, pp. 1–12 (2014)
13. Zanella, A., Bui, N., Castellani, A.P., Vangelista, L., Zorzi, M.: Internet of things for smart cities. IEEE Internet Things J. **1**, 22–32 (2014)
14. Grieco, L.A., Alaya, M.B., Monteil, T., Drira, K.K.: Architecting information centric ETSI-M2 M systems. In: IEEE PerCom (2014)
15. Anderson, J., Rainie, L.: The internet of things will thrive by 2025, Pew research internet project (2014). http://www.pewinternet.org/2014/05/14/internet-of-things/
16. Yan, Z., Zhang, P., Vasilakos, A.V.: A survey on trust management for internet of things. J. Netw. Comput. Appl. **42**, 120–134 (2014)
17. Piro, G., Boggia, G., Grieco, L.A.: A standard compliant security framework for IEEE 802.15.4 networks. In: Proceedings of IEEE World Forum on Internet of Things (WF-IoT), Seoul, South Korea, pp. 27–30 (2014)
18. Lee, J.-Y., Lin, W.-C., Huang, Y.-H.: A lightweight authentication protocol for internet of things. In: 2014 International Symposium on Next-Generation Electronics, ISNE 2014, Kwei-Shan, pp. 1–2 (2014)
19. Turkanovi, M., Brumen, B., Hlbl, M.: A novel user authentication and key agreement scheme for heterogeneous ad hoc wireless sensor networks, based on the internet of things notion. Ad Hoc Netw. **20**, 96–112 (2014)
20. Ye, N., Zhu, Y., Wang, R.-C.B., Malekian, R., Lin, Q.-M.: An efficient authentication and access control scheme for perception layer of internet of things. Appl. Math. Inf. Sci. **8**(4), 1617–1624 (2014)
21. Cherkaoui, A., Bossuet, L., Seitz, L., Selander, G., Borgaonkar, R.: New paradigms for access control in constrained environments. In: 2014 9th International Symposium on Reconfigurable and Communication-Centric Systems-on-Chip (ReCoSoC), Montpellier, pp. 1–4 (2014)
22. Peng, L.B., Ru-chuan, W.B., Xiao-yu, S., Long, C.: Privacy protection based on key-changed mutual authentication protocol in internet of things. Commun. Comput. Inf. Sci. **418**, 345–355 (2014)
23. Ukil, A., Bandyopadhyay, S., Pal, A.: IoT-privacy: to be private or not to be private. In: Proceedings—IEEE INFOCOM, Toronto, ON, pp. 123–124 (2014)
24. Sicari, S., Cappiello, C., Pellegrini, F.D., Miorandi, D., Coen-Porisini, A.: A security-and quality-aware system architecture for internet of things. Inf. Syst. Front. **18**, 1–13 (2014)
25. Tormo, G.D., Marmol, F.G., Perez, G.M.: Dynamic and flexible selection of a reputation mechanism for heterogeneous environments. Futur. Gener. Comput. Syst. **49**, 113–124 (2014)
26. Gu, L., Wang, J., Sun, B.B.: Trust management mechanism for internet of things. China Commun. **11**(2), 148–156 (2014)
27. Liu, Y.-B., Gong, X.-H., Feng, Y.-F.: Trust systembased on node behavior detection in internet of things. Tongxin Xuebao/J. Commun. **35**(5), 8–15 (2014)

28. Singh, J., Bacon, J., Eyers, D.: Policy enforcement within emerging distributed, event-based systems. In: DEBS 2014—Proceedings of the 8th ACM International Conference on Distributed Event-Based Systems, pp. 246–255 (2014)
29. Neisse, R., Steri, G., Baldini, G.: Enforcement of security policy rules for the internet of things. In: Proceedings of IEEE WiMob, Larnaca, Cyprus, pp. 120–127 (2014)
30. Ferreira, H., De Sousa Jr, R., De Deus, F., Canedo, E.: Proposal of a secure, deployable and transparent middleware for internet of things. In: Iberian Conference on Information Systems and Technologies. CISTI, Barcelona, pp. 1–4 (2014)
31. Niu, B., Zhu, X., Chi, H., Li, H.: Privacy and authentication protocol for mobile RFID systems. Wireless Pers. Commun. **77**(3), 1713–1731 (2014)
32. Jeong, Y.-S., Lee, J., Lee, J.-B., Jung, J.-J., Park, J.: An efficient and secure m-IPS scheme of mobile devices for human-centric computing. J. Appl. Math. **2014**, 1–8 (2014)
33. Geng, J., Xiong, X.: Research on mobile information access based on internet of things. Appl. Mech. Mater. **539**, 460–463 (2014)
34. Kubler, S., Frmling, K., Buda, A.: A standardized approach to deal with firewall and mobility policies in the IoT. Pervasive Mobile Comput. **20**, 100–114 (2014)
35. Daubert, J., Wiesmaier, A., Kikiras, P.: A view on privacy & trust in IoT. In: IOT/CPS-Security Workshop, IEEE International Conference on Communications, ICC 2015, London, GB, June 08–12, 2015, page to appear. IEEE (2015)
36. Bhattacharjya, A., Zhong, X., Wang, J.: Strong, efficient and reliable personal messaging peer to peer architecture based on Hybrid RSA. In: Proceedings of the International Conference on Internet of Things and Cloud Computing (ICC 2016) ISBN 978-1-4503-4063-2/16/03. The Møller Centre-Churchill College, Cambridge (2016). https://doi.org/10.1145/2896387.2896431
37. Bhattacharjya, A., Zhong, X., Wang, J.: An end to end users two way authenticated double encrypted messaging scheme based on Hybrid RSA for the Future Internet Architectures. Int. J. Info Comput. Secur. **10**, 63–79 (2017). https://doi.org/10.1504/IJICS.2018.10005506
38. Bhattacharjya, A., Zhong, X., Wang, J., et al.: On mapping of address and port using translation (MAP-T). Abstract Published in Int. J. Info Comput. Secur. **11**(3), 214–232 (2019)
39. Sicari, S., Rizzardi, A., Cappiello, C., Coen-Porisini, A.: A NFP model for internet of things applications. In: Proceedings of IEEE WiMob, Larnaca, Cyprus, pp. 164–171 (2014)
40. Wang, X., Zhang, J., Schooler, E., Ion, M.: Performance evaluation of attribute-based encryption: toward data privacy in the IoT. In: 2014 IEEE International Conference on Communications, ICC 2014, Sydney, NSW, pp. 725–730 (2014)
41. Su, J., Cao, D., Zhao, B., Wang, X., You, I.: ePASS: an expressive attribute-based signature scheme with privacy and an unforgeability guarantee for the internet of things. Futur. Gener. Comput. Syst. **33**, 11–18 (2014)
42. Gòmez-Goiri, A., Orduna, P., Diego, J., de Ipina, D.L.: Otsopack: lightweight framework for interoperable ambient intelligence applications. Comput. Hum. Behav. **30**, 460–467 (2014)
43. Colistra, G., Pilloni, V., Atzori, L.: The problem of task allocation in the internet of things and the consensus-based approach. Comput. Netw. **73**, 98–111 (2014)
44. Wang, Y., Qiao, M., Tang, H., Pei, H.: Middleware development method for internet of things. Liaoning Gongcheng Jishu Daxue Xuebao (Ziran Kexue Ban)/J Liaoning Tech Univ (Nat Sci Ed) **33**(5), 675–678 (2014)
45. Bhattacharjya, A., Zhong, X., Wang, J.: HYBRID RSA based highly efficient, reliable and strong personal Full Mesh Networked messaging scheme. Int. J. Info Comput. Secur. **10**(4), 418–436 (2018)
46. Bhattacharjya, A., Zhong, X., Wang, J., et al.: Security challenges and concerns of internet of things (IoT), Cyber-Physical System: Architecture, Security and Application. EAI/Springer Innovations in Communications and Computing, 153–185 (2019)
47. Bhattacharjya, A., Zhong, X., Wang, J., et al.: Secure IoT Structural design for Smart Homes, Smart Cities Cybersecurity and Privacy. 187–201, Elsevier. http://www.sciencedirect.com/science/article/pii/B9780128150320000135

48. Bhattacharjya, A., Zhong, X., Wang, J., Xing, L.: An efficient and four-layered authenticated secure Hybrid RSA (SHRSA) messaging scheme. IEEE Access, 7, 30487–30506 (2019). Digital Object Identifier https://doi.org/10.1109/ACCESS.2019.2900300

Aniruddha Bhattacharjya is with the Department of Electronic Engineering, Tsinghua University, Beijing, China, as a Chinese Government Ph.D. scholar. His research interests are cryptography, network security, RFID-based architectures and middleware, security in fixed and wireless Networks, applications of cryptography, and IoT security. He has received the ICDCN 2010, Ph.D. Forum Fellowship. He achieved the best paper award in ACM ICC 2016 in Cambridge University, UK. Since 2012, he has been working as an IEEE mentor and ACM faculty sponsor. He is a member of 34 IEEE societies and various IEEE technical committees. He has published 35 papers as well as 1 Chinese innovation patent is filed.

Xiaofeng Zhong received his Ph.D. in Information and Communication Systems from Tsinghua University in 2005. He is an Associate Professor in the Department of Electronic Engineering at Tsinghua University. He performs research in the field of mobile networks, including users' behaviors and traffic model analyses, MAC and network protocol design, and resource management optimization. He has published more than 30 papers and holds 7 patents.

Jing Wang received his BS and MS degree in Electronic Engineering from Tsinghua University, Beijing, China in 1983 and 1986, respectively. He has worked as a Faculty member in Tsinghua University since 1986. He is currently a Professor at the School of Information Science and Technology. He also serves as the Vice Director of the Tsinghua National Laboratory for Information Science and Technology. His research interests are in the area of wireless communications, including transmission and networking technologies of 4G/5G. He has published more than 150 conference and journal papers.

Xing Li received the B.S. degree in radio electronics from Tsinghua University, Beijing, China, in 1982, and the M.S. and Ph.D. degrees in electrical engineering from Drexel University, Philadelphia, PA, USA, in 1985 and 1989, respectively. He is currently a Professor with the Electronic Engineering Department, Tsinghua University. His research activities and interests include statistical signal processing, multimedia communication, and computer networks. He has published more than 300 papers in his research areas. He is a Deputy Director of the China Education and Research Network (CERNET) Center and a Member of the Technical Board of the CERNET Project. He was a Member of Communication Expert Committee of the China National "863" High Technology Project. He is a Formal Chairman of the Asia Pacific Networking Group and a Formal Member of the executive council of the Asia Pacific Network Information Center.

Some Computational Considerations for Kernel-Based Support Vector Machine

Mohsen Esmaeilbeigi, Alireza Daneshkhah and Omid Chatrabgoun

Abstract Sometimes healthcare perspectives in communications technologies require data mining, especially classification as a supervised learning. Support vector machines (SVMs) are considered as efficient supervised learning approaches for classification due to their robustness against several types of model misspecifications and outliers. Kernel-based SVMs are known to be more flexible tools for a wide range of supervised learning tasks and can efficiently handle non-linear relationship between input variables and outputs (or labels). They are more robust with respect to the aforementioned model misspecifications, and also more accurate in the sense that the root-mean-square error computed by fitting the kernel-based SVMs is considerably smaller than the one computed by fitting the standard/linear SVMs. However, the choice of kernel type and particularity kernel's parameters could have significant impact on the classification accuracy and other supervised learning tasks required in network security, Internet of things, cybersecurity, etc. One of the findings of this study is that larger kernel parameter(s) would encourage SVMs with more localities and vice versa. This chapter provides some results on the effect of the kernel parameter on the kernel-based SVM classification. We thus first examine the effect of these parameters on the classification results using the kernel-based SVM, and then specify the optimal value of these parameters using cross-validation (CV) technique.

Keywords Classification · Cross-validation · Support vector machine (SVM) · Kernel-based SVM

M. Esmaeilbeigi · O. Chatrabgoun
Department of Statistics, Faculty of Mathematical Sciences and Statistics, Malayer University, Malayer, Iran
e-mail: m.esmaeilbeigi@malayeru.ac.ir

O. Chatrabgoun
e-mail: o.chatrabgoun@malayeru.ac.ir

A. Daneshkhah (✉)
Faculty of Engineering, Environment and Computing, Coventry University, Coventry CV1 2JH, UK
e-mail: ac5916@coventry.ac.uk; ali.daneshkhah@coventry.ac.uk

M. Farsi et al. (eds.), *Digital Twin Technologies and Smart Cities*, Internet of Things, https://doi.org/10.1007/978-3-030-18732-3_10

1 Introduction

In the recent years, data mining or the extraction of knowledge from the available data has received considerable attention in many fields including communication technologies [1]. Using data mining methods which can be generally divided into supervised and unsupervised learning, pivotal features/information can be retrieved from the existing data without any presumptions that are required [2]. Roughly speaking, supervised learning is known as classification in its discrete form, and regression modelling in the continuous form. The classification has various forms in which the binary classification is one of the widely used methods. There are several tools available for this type of classification, including Gaussian processes (GP) [3, 4], neural networks [5, 6], or support vector machine (SVM) [7] which has recently received significant attention. The main aim of the first former methods is to reduce the classification error, but the main purpose of the SVM is to reduce the operational risk in classification. Various versions of the SVM have been recently proposed and used in the wide range of applications as addressed in [8, 9].

It has been a common practice to classify the existing data as a hyperplane using a linear SVM [10]. However, the classification efficiency of linear SVM is highly dependent on the available data. The linear SVMs would not usually perform well on data that is not linearly separable. Another major practical problem with this SVM is the high algorithmic complexity and extensive memory requirements [11]. In these situations, the kernel-based SVM is recommended to be considered due to the flexibility of the non-linear kernel which allows to develop a classifier using the feature space than data (input) space. In other words, instead of using data space for classification, we can use a property in the data, such as distance from each other, or function of this distance, known as *feature space*. Kernel-based SVM is a more useful approach to deal with non-linear classification based on a linear discriminant function in a high-dimensional (kernel) space [8, 12]. It should be noted that the capacity in the kernel-based SVM generally refers to the useful characterisations of the kernels such as the possibility of interacting with high-dimensional data and the possibility of encountering with non-linear data. In addition, the existence of some customizable parameters in the kernels makes the kernel-based SVM more flexible in classification (see [13] for further details). In other words, the parameters in kernels are strongly influencing the classification results using the SVM approach.

In this chapter, we first examine the effect of these parameters on the classification results with kernel-based SVM, and we then find the optimal values of these parameters using cross-validation (CV) technique which is widely used as a data-driven method to predict optimal values of model parameters.

2 SVM Classification

In this section, we first briefly introduce some preliminary results of the SVM where the main focus is to study the advantages and drawbacks of the linear SVM.

2.1 Preliminaries

Suppose that we are given a set of training data $\{(x_i, y_i)|i = 1, \ldots, N\}$ with measurements $x_i \in \mathbb{R}^d$ and data values in the form of labels $y_i \in \{1, +1\}$. The standard (binary) classification problem consists of finding a predictor that will allow us to assign an appropriate label, either 1 or $+1$, to a future measurement x. Such a predictor might be of the form $\text{sign}(h(x))$, where h depends on dimension d in our given measurements denoted by a line

$$w_1x_1 + w_2x_2 + b = 0,$$

plane

$$w_1x_1 + w_2x_2 + w_3x_3 + b = 0,$$

or hyperplane

$$\sum_{i=1}^{n} w_ix_i + b = 0,$$

which are separable. In the general and matrix form $w^Tx + b = 0$, such that the weights w serve as the unit normal vector to the hyperplane and b denotes the interception or bias term. In fact, the rule of classification is that

$$\text{if} \quad y_i = 1 \Rightarrow \; w^Tx_i + b > 1, \tag{1}$$

$$\text{if} \quad y_iy_i = -1 \Rightarrow \; w^Tx_i + b < -1. \tag{2}$$

In the simplest case, our predictor is given by $\text{sign}(h(x))$, where h denotes a hyperplane, directly in input space, of the form

$$h(x) = x^Tw + b, \quad x \in \mathbb{R}^d,$$

that separates the measurements with label -1 from those with label $+1$. The weight w and the bias b can be determined by maximizing the margin or gap to both sides of this hyperplane (see also [1]). Since the size of this margin equals to $\frac{1}{||w||}$, and we would ideally like to have this margin as large as possible, we want to minimize $||w||$, the norm of the coefficients of h. Thus, by remembering $h(x) = x^Tw + b$, we would have an unconstrained minimization problem of the form min $||w||$, or

$$\min \ \frac{1}{2}||w||^2 = \frac{1}{2}w^T w,$$

by combining both conditions (1) and (2), we can create the following constraint:

$$y_i(w^T x_i + b) > 1, \tag{3}$$

such that we are faced with a constrained optimization problem. Instead of this formulation, the following constrained optimization with slack variables ε_i is more common since it also allows us to deal with the case where the given measurements are not perfectly separable by h:

$$\min \ \frac{1}{2}w^T w + C\sum_{i=1}^{N} \varepsilon_i,$$

such that with slack variables ε_i, the constraint (3) will be changed to

$$y_i(w^T x_i + b) > 1 - \varepsilon_i.$$

The considered formulation can be derived via Lagrange multipliers α_i and is known in the SVM literature as the primal problem, i.e.,

$$L_P = \frac{1}{2}w^T w + C\sum_{i=1}^{N} \varepsilon_i - \sum_{i=1}^{N} \alpha_i \left[y_i(w^t x_i + b) - 1 + \varepsilon_i\right].$$

The corresponding (SVM) primal problem can be changed to dual problem which follows from setting the w-gradient of the primal Lagrange multiplier functional equals to zero and is of the form

$$L_D = -\frac{1}{2}\sum_{i=1}^{N}\sum_{j=1}^{N} \alpha_i \alpha_j y_i y_j x_i^T x_j^T + \sum_{i=1}^{N} \alpha_i,$$

subject to $\sum_{i=1}^{N} \alpha_i y_i = 0$, and $0 \leq \alpha_i \leq C$, where C is known as a box constraint, and

$$w = \sum_{i=1}^{N} \alpha_i y_i x_i.$$

The bias b_i is given by $b_i = y_i x_i^T w$, for any i and the optimal $\alpha_i > 0$. For stability purposes, we actually consider all qualifying indices and find b using the mean. Note that the box constraint C is a free parameter which needs to be either set by the user or determined by additional parameter optimization methods such as cross-validation.

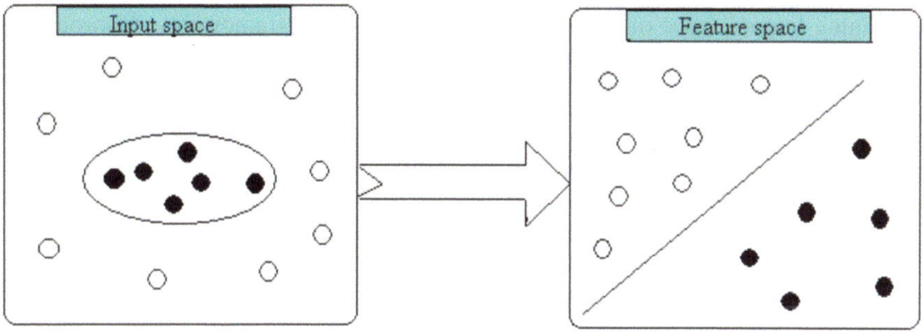

Fig. 1 Data not linear separable in input space, but they are in the feature space

2.2 *Kernel-Based Classification*

Sometimes, it is not possible to classify existing data with one hyperplane and separate them using a linear SVM. For this reason, we consider a feature space in place of the data itself or input space, and try to separate data in feature space by linear SVM or hyperplane. For example, the feature space can be the distance of data from each other or a function of this distance. If we show the feature space of the measurements with $\phi(x)$, as shown in Fig. 1, they are not linearly separable with the input space, but in the feature space they are. Note that this feature space is potentially infinite-dimensional and therefore offers much more flexibility for separating the data than the finite-dimensional input space. This fact has a theoretical foundation in the form of Cover's theorem [14], which ensures that data which cannot be separated by a hyperplane in input space most likely will be linearly separable after being transformed to feature space by a suitable feature map. Thus, support vector machines—in terms of feature space, in particular—are a good tool to use in order to tackle intricate data classification problems.

The algorithms for non-linear classification are now more or less the same as before; simply replace the measurements x_i in input space by their features ϕ_{x_i} in feature space. The separating hyperplane can be expressed in the form

$$h(x) = \phi_x^T w + b = 0, \quad x \in \mathbb{R}^d,$$

and the dual problem for SVM classification using the transformed input data is given by

$$\begin{aligned} \min \ & \frac{1}{2} \sum_{i=1}^{N} \sum_{j=1}^{N} \alpha_i \alpha_j y_i y_j \phi(x_i)^T \phi(x_j) - \sum_{i=1}^{N} \alpha_i \\ \text{s.t} \ & \sum_{i=1}^{N} \alpha_i y_i = 0 \quad \& \quad 0 \le \alpha_i \le C. \end{aligned}$$

Obtaining feature space of the data is possible within the framework of reproducing kernel Hilbert space (RKHS). In other words, the mapping of the introduced feature is considered as $\phi : \omega \longrightarrow H_K(\omega)$ under the map

$$X \longmapsto \phi_X = K(.,X).$$

The data set is transferred from space ω to feature space $H_K(\omega)$, where $H_K(\omega)$ is RKHS corresponded to the kernel K (see [13] for further details). Regarding RKHS characterizations, the inner product in the feature space, $H_K(\omega)$, using the kernel K is simply possible, namely,

$$K(X, Z) = \phi_X^T \phi_Z.$$

Now, with the help of the above relation, a more general form of the quadratic optimization problem in the feature space of SVM is as follows:

$$\min \frac{1}{2}\sum_{i=1}^{N}\sum_{j=1}^{N}\alpha_i\alpha_j y_i y_j k(x_i, x_j) - \sum_{i=1}^{N}\alpha_i$$

$$\text{s.t} \sum_{i=1}^{N}\alpha_i y_i = 0 \quad \& \quad 0 \le \alpha_i \le C.$$

Given the important role of the kernels, the above approach is named as a kernel-based SVM and provides the possibility of non-linear classification of data. It should be noted that the capacity in the kernel-based SVM generally refers to the useful features of the kernels such as the possibility of interacting with high-dimensional data and the possibility of encountering data that are classified nonlinearly. Additionally, the existence of some customizable parameters in the kernels makes the kernel-based SVM more flexible. Generally, the use of known kernels in the kernel-based SVM is more common than other kernels. These kernels are the polynomial kernel, the Gaussian kernel, and the sigmoid kernel, as their structures are shown in Table 1. Each of these kernels has a shape parameter that makes them more flexible. Moreover, kernels may be defined via the feature map (instead of in closed form), and this feature map can be picked depending on the specific application at hand (e.g., as a string kernel for text mining). The ubiquitous Gaussian kernel does not need much of an introduction, although it might be curious to note that it seems to have been first mentioned in its role as a probability density function in the context of least squares approximation and maximum likelihood estimation. The Gaussian kernel is infinitely differentiable and positive definite. It will be used in this chapter for the kernel-based SVM. In the literature on statistics and machine learning, the Gaussian kernel is sometimes referred to as the squared exponential kernel, where

$$w = \sum_{i=1}^{N}\alpha_i y_i K(x, x_i),$$

Table 1 Polynomial, Gaussian, and sigmoid kernel structure to use in kernel-based SVM

Kernel	Structure $(K(X, Z))$
Gaussian	$e^{\varepsilon\|\|x-z\|\|}$
Polynomial	$(1 + x^T z)^\varepsilon$
Sigmoid	$tanh(1 + \varepsilon x^T z)$

the classifier $h(x)$ is now given by

$$\text{sign}(h(x)) = sign(\sum_{i=1}^{N} \alpha_i y_i k(x_i, x) + b),$$

where b_i is obtained as before, i.e., $b_i = y_i - \sum_{j=1}^{N} \alpha_j y_j k(x_j, x_i)$ with i denoting the index of an α_i which is strictly between 0 and C. For stability purposes, we can again average over all such candidates. For positive definite kernels, it is also possible to formulate the separating hyperplane without the bias term b (see, e.g., [15]).

We note that the hyperplane will be linear only in feature space (which we usually have no concrete knowledge of). In the input space, the data will be separated by a non-linear manifold. Moreover, the representation of this manifold is sparse in the sense that not all basic functions are needed to specify it. In fact, only those centres x_j whose corresponding $_j$ are non-zero define meaningful basis functions. These special centres are referred to as support vectors.

Also, since the decision boundary can be expressed in terms of a limited number of support vectors, i.e., it has a sparse representation, learning is possible in very high-dimensional input spaces (see, e.g., [8]). Moreover, SVMs are robust against several types of model violations and outliers, and they are computationally efficient, e.g., using the sequential minimal optimization (SMO) algorithm of Platt [16] to perform the quadratic optimization task required for classification as well as regression. Another way to make SVMs perform more efficiently is to consider a low-rank representation for the kernel as in [17].

Matlab's statistics and machine learning toolbox provides outstanding functions and documentation for training and predicting with support vector machines which will be followed in the next section.

3 Numerical Results

This example uses a pattern which is not linearly separable and attempts to classify it: population 1 (denoted by blue $\boxdot$ and green $+$) has centre points at $\{(0, 1), (1, 0), (2, 1)\}$ and population 2 (denoted by blue $\bigcirc$ and red $\times$) has centre points at $\{(0, 0), (1, 1), (2, 0)\}$. Test points are chosen from a normal distribution from those populations and training data is generated from the test points. These distribution

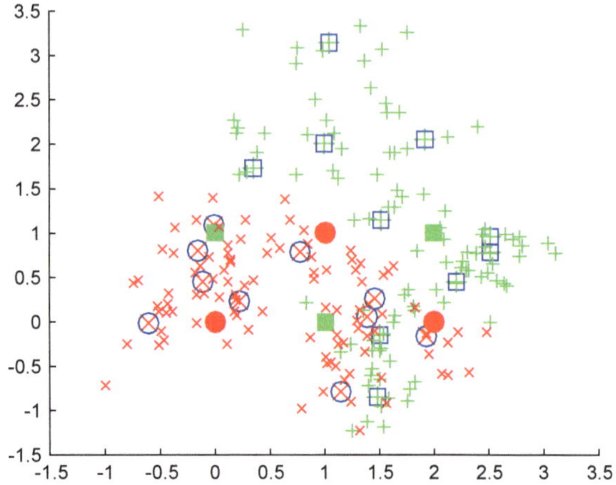

Fig. 2 Scatter plot of input data and test points for population 1 versus population 2

centre points (filled green ⊡ and red ◯), test points (large green + and red ×) and training points (small green + and red ×) are on display in Fig. 2. This pattern is not linearly separable, and this fact is evident in Fig. 2.

The decision contours show the impact of the various kernel parametrizations on the SVM. Figure 3a, b alternately fixes C and ε values and allows the other parameter to vary to show the impact on the SVM. Clearly, the choice of ε plays a significant role, where larger ε encourages an SVM with more locality and smaller ε encourages less localized influence; this matches the standard localization behaviour for Gaussian kernel in an interpolation setting. When $\varepsilon = 1$ is fixed and different C values are considered, a similar impact occurs. It seems that smaller C values produce a less active decision contour, whereas large C encourages more local fluctuations. Now, we consider fixing $C = 0.6$ with a variety of ε values. Figure 4a, b shows the number of missed classifications (out of 20 possible) as well as the margin $||w||^{-1}$ and the required number of support vectors, respectively. The margin does not appear to be a useful guide for determining an optimal ε value, as the margin grows unboundedly as $\varepsilon \longrightarrow 0$; on the other hand, for a very large ε, this example is perfectly classified. Minimizing the number of support vectors is optimal from a computational standpoint, and also seems to suggest a viable region for predictions. When we fix $\varepsilon = 0.01$ and consider a range of C values, as illustrated in Fig. 5a, b, the number of missed classification will grow. Good prediction results seem to occur for smaller C values, and very large values of C which require few support vectors. It is worth noting that larger C values require more computing time to solve the quadratic programme because a larger search space is under consideration. The numerical result demonstrates that finding an optimal SVM parametrization using either the margin or the number of support vectors is not always a useful strategy. A more common technique in the machine learning community is to use cross-validation (CV).

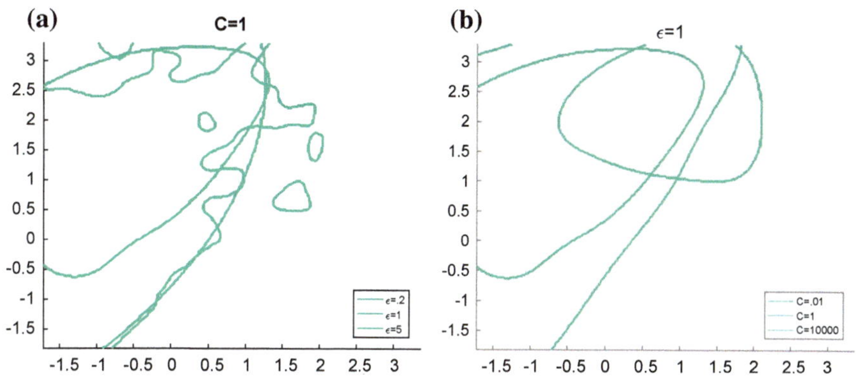

Fig. 3 Fixes C (**a**) and ε (**b**) values and allows the other parameter to vary to show the impact on the SVM

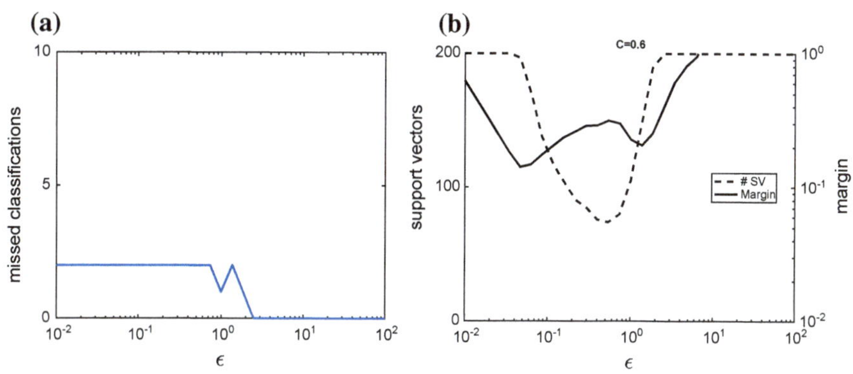

Fig. 4 Fixes $C = 0.6$ with a variety of ε which shows the number of missed classifications (**a**) as well as the margin and the required number of support vectors (**b**)

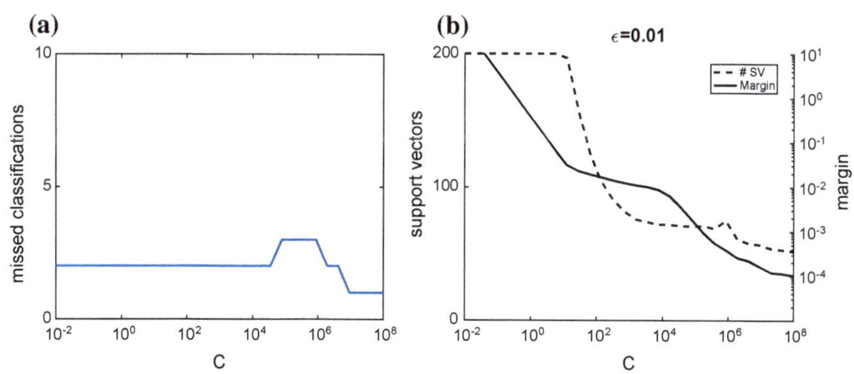

Fig. 5 Fixes $\varepsilon = 0.01$ with a variety of C which shows the number of missed classifications (**a**) as well as the margin and the required number of support vectors (**b**)

3.1 Kernel-Based SVM with Cross-Validation

Cross-validation is a popular technique in statistics which uses the given data (instead of the usually unknown solution) to predict optimal values of model parameters for data fitting. The main idea is to split the data into a training set τ and a validation set ν and to then use some form of error norm obtained by gauging the accuracy of the fit built from information on the training set at points in the validation set.

Cross-validation (CV) is an especially popular version of cross-validation and corresponds to using a training set consisting of all but some of the data points, which in turn are the sole member of the validation set. In the context of kernel methods, CV appeared in papers such as [18–20]. Often times, cross-validation is conducted in one of two ways:

Leave-one-out cross-validation (LOOCV): All the data except a single point is used to compute in kernel-based SVM classification, and the residual is judged at that point. In this setting, $V = \{\nu^{(1)}, \nu^{V(2)}, \ldots, \nu^{V(N)}\} = \{x_1, x_2, \ldots, x_N\}$ and the errors at each of those points are added up to find total error. As explained in [21], it is most

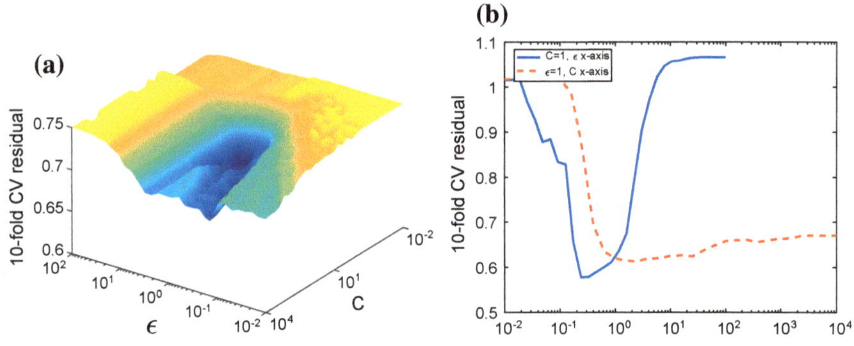

Fig. 6 Fixes $\varepsilon = 0.01$ with a variety of C which shows the number of missed classifications (**a**) as well as the margin and the required number of support vectors (**b**)

Fig. 7 The number of missed classification items in terms of two parameters ε and C

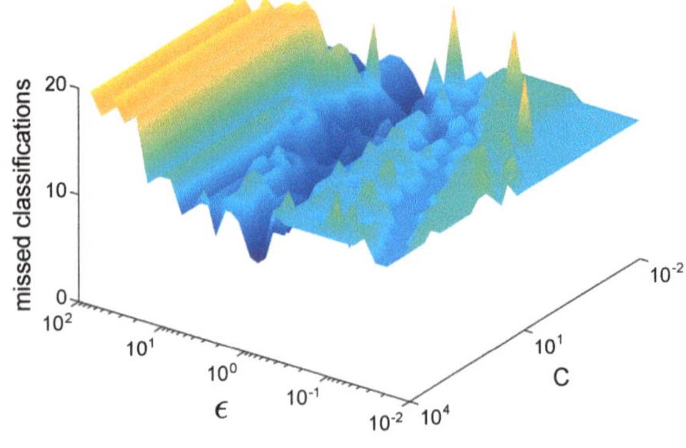

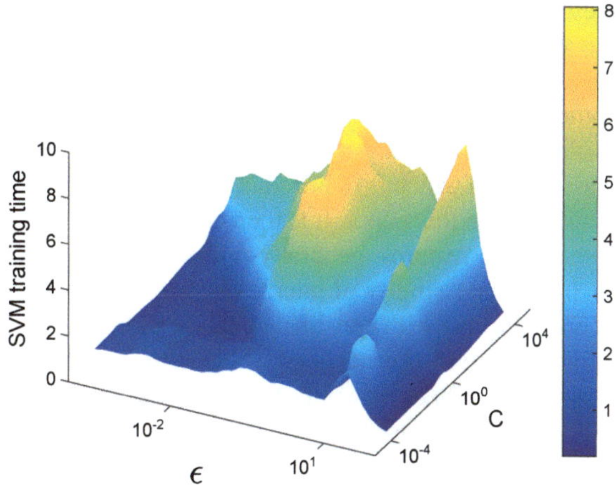

Fig. 8 Training time for classification in terms of two parameters ε and C

likely the preferred form to compute. Thus, the entire LOOCV computation can be performed with little overhead compared to the computation of the classification.

Leave-half-out cross-validation: Half of the data is omitted to create a classification and the residual is judged on the other half; then the process is flipped and both results are combined to compute total error. In this setting, $V = \{\nu^{(1)}, \nu^{(2)}\}$ and $|\nu^{(1)}| = |\nu^{(1)}|$, or as close as possible. Sometimes, instead of leave-half-out cross-validation, some part of the data preserve for cross-validation. This part of data is called fold.

Here, a 10-fold cross-validation scheme is used to measure the effectiveness of each of the ε and C parameters. Figure 6a shows the associated cross-validation residuals in terms of two parameters ε and C. Figure 6b also shows the associated cross-validation residuals with one parameter fixed and the other varied over $[10^{-2}, 10^{2}]$. It can be also demonstrated that there is an optimal region for the 10-fold cross-validation residual, where decreases in ε are matched by increases in C. Figure 7 shows the number of missed classification items in terms of two parameters ε and C. Figure 8 also illustrates the training time for classification in terms of two parameters ε and C.

4 Conclusion

In this paper, we have studied the advantages and drawbacks of various SVMs, including linear and kernel-based methods, for the classification of some applications in healthcare and communication technologies. Since SVMs are known to be robust against several types of model misspecification and outliers, the SVM-based classification approaches should also be robust against these features. We illustrated that the linear SVM could be useful in classifying the data with specific features, and they are most efficient to be generally used for any dataset. An alternative method is the kernel-based SVM which is constructed using the feature space instead of data (input) space. In other words, instead of using data space for classification, we use some nice properties in the data, such as distance from each other, or function of this distance (i.e., feature space). The kernel-based SVM approach allows non-linear classification which is required in the most complex applications, including the ones considered in this paper. It is then demonstrated that the kernel-based SVM possess some useful features, generally refers to the useful characterizations of the kernels such as the possibility of interacting with high-dimensional data and the possibility of encountering with non-linear data. In addition, the existence of some customizable parameters in the kernels makes the kernel-based SVM more flexible in non-linear classification. In other words, the produced error using linear SVMs has been reduced using the kernel-based SVMs. The choice of kernel parameter obviously plays a significant role, when larger kernel parameter encouraged SVMs with more localities and vice versa. This paper has provided some results on the effect of the parameter of the kernel-based SVM classification. In fact, we have examined the effect of these parameters on the classification performance using the kernel-based SVM by appropriately determining the optimal value of these parameters using the presented cross-validation (CV) technique.

References

1. Hastie, T., Tibshirani, R., Friedman, J.: Elements of Statistical Learning: Data Mining, Inference, and Prediction. Springer Series in Statistics, 2nd edn. Springer, New York (2009)
2. Han, J., Kamber, M., Pei, J.: Data Mining Concepts and Techniques, 3rd edn. Elsevier Inc., Waltham, USA (2012)
3. Nickisch, H., Rasmussen, C.E.: Approximations for binary gaussian process classification. J. Mach. Learn. Res. **9**, 2035–2078 (2008)
4. Rasmussen, C.E., Williams, C.K.I.: Gaussian Processes for Machine Learning. MIT Press (2006)
5. Arajo, R.A., Oliveira, A.L.I., Meira, S.: A morphological neural network for binary classification problems. Eng. Appl. Artif. Intell. **65**, 12–28 (2017)
6. Qian, G., Zhang, L.: A simple feedforward convolutional conceptor neural network for classification. Appl. Soft Comput. **70**, 1034–1041 (2018)
7. Wang, L.: Support Vector Machines: Theory and Applications. Springer, Berlin (2005)
8. Steinwart, I., Christmann, A.: Support Vector Machines, Information Sciences and Statistics. Springer, New York (2008)
9. Vapnik, V.: The Nature of Statistical Learning Theory. Springer, New York (2013)
10. Burges, C.J.: A tutorial on support vector machines for pattern recognition. Data Min. Knowl. Discov **2**(2), 121–167 (1998)
11. Suykens, J.A.: Advances in Learning Theory: Methods, Models, and Applications, vol. 190. IOS Press (2003)
12. Scholkopf, B., Smola, A.J.: Learning with Kernels: Support Vector Machines, Regularization, Optimization, and Beyond. MIT Press, Cambridge (2002)
13. Fasshauer, G., McCourt, M.: Kernel-Based Approximation Method using Matlab. World Scientific Publishing (2016)
14. Cover, T.M.: Geometrical and statistical properties of systems of linear inequalities with applications in pattern recognition. IEEE Trans. Electron. **14**, 326–334 (1965)

15. Poggio, T., Mukherjee, S., Rifkin, R., Rakhlin, A., Verri, A.: Technical report, MIT AI Memo 2001-011 (2001)
16. Platt, J.C.: Fast training of support vector machines using sequential minimal optimization. In: Scholkopf, B., Burges, C.J.C., Smola, A.J. (eds.) Advances in Kernel Methods, pp. 185–208. MIT Press, Cambridge, MA (1999)
17. Fine, S., Scheinberg, K.: Efficient SVM training using low-rank kernel representations. J. Mach. Learn. Res. **2**, 243–264 (2002)
18. Hickernell, F.J., Hon, Y.C.: Radial basis function approximations as smoothing splines. Appl. Math. Comput. **102**(1), 1–24 (1999)
19. Rippa, S.: An algorithm for selecting a good value for the parameter c in radial basis function interpolation. Adv. Comput. Math. **11**(2–3), 193–210 (1999)
20. Fasshauer, G.E., Zhang, J.G.: On choosing "optimal" shape parameters for RBF approximation. Numer. Algorithms **45**(1–4), 345–368 (2007)
21. Fasshauer, G.E.: Meshfree Approximation Methods with Matlab, Interdisciplinary Mathematical Sciences, vol. 6. World Scientific Publishing Co., Singapore (2007)

A Secure Hybrid RSA (SHRSA)-based Lightweight and Efficient Personal Messaging Communication Protocol

Aniruddha Bhattacharjya, Xiaofeng Zhong, Jing Wang and Xing Li

Abstract Balancing efficiency, privacy, and security along with strong authentication in the End-to-End (E2E) communication is a burning issue in personal messaging. Rivest–Shamir–Adleman (RSA) algorithm is an omnipresent cryptographic approach, so here we have implemented a Secure Hybrid RSA (SHRSA)-based lightweight and efficient personal messaging communication protocol for E2E secure, authenticated, and efficient messaging. The SHRSA decryption is much more secure and efficient than RSA and Chinese Remainder Theorem (CRT)-RSA. It is protecting the messaging scheme users with a perfect privacy. The SHRSA cipher's communication protocol is resolving many RSA-related issues. Full mesh networked personal messaging communication protocol ensures E2E encryption for all peers. The testing results of the personal messaging communication protocol have proved that this protocol is an efficient and secure personal messaging communication protocol. Also, it occupies very less memory and very less CPU than RSA and CRT-RSA. So high security, decryption efficiency with less memory and less CPU occupancy features make this secure message communication protocol much relevant to the era of Internet of Everything (IoE). Also, it is relevant to other secure and authentic message communications.

Keywords Secure hybrid RSA (SHRSA) · E2E encryption · Multilayered communication protocol · SHRSA encryption · SHRSA decryption · Decryption efficiency, Internet of Everything (IoE)

Jing Wang died before publication of this work was completed.

A. Bhattacharjya (✉) · X. Zhong · J. Wang · X. Li
Department of Electronic Engineering, Beijing National Research Center for Information Science and Technology, Tsinghua University, Beijing, China
e-mail: li-an15@mails.tsinghua.edu.cn

X. Zhong
e-mail: zhongxf@tsinghua.edu.cn

J. Wang
e-mail: wangj@tsinghua.edu.cn

X. Li
e-mail: xing@cernet.edu

M. Farsi et al. (eds.), *Digital Twin Technologies and Smart Cities*,
Internet of Things, https://doi.org/10.1007/978-3-030-18732-3_11

1 Introduction

Cryptography which is the science of encryption plays a dominant role in providing security to Automated Teller Machine (ATM) cards, sending or receiving private emails, electronic commerce, digital signature, mobile phone communications, transaction processing, securing computer from unauthorized access, and also related to many aspects of the day-to-day lives. In the scenario of personal messaging, we need E2E encryption, which should protect the privacy with authentication, but it also should balance the efficiency. RSA cipher is suffering from various backlogs in this era, particularly on the personal messaging scenario. Therefore, it is vivid that, a secure, strongly authenticated, reliable messaging scheme with strong E2E protection to users and the messages, which will be also much faster, is an essential need these days. Therefore, efficiency plays a pivot role in the designing of secure communication protocol. One of the good solutions can be variants of RSA [1–18], which can make a new hybrid RSA messaging communication protocol, for balancing the efficiency with strong security, privacy, reliability, and authentication. It is well understood that balancing strong security, privacy, reliability, and authentication with high efficiency is challenging. There are several variants of RSA algorithm proposed in the past. The proposed secure communication protocol should be protective to privacy strongly, should replace some of the backlogs of RSA, and should work purely peer-to-peer enabling end-to-end encryption.

We know a recent term "Digital Twin". The idea of a virtual, digital alike to the physical product or the Digital Twin was introduced in 2003 by Dr. Michael Grieves. It was termed in Virtually Perfect: Driving Innovative and Lean Products through Product Lifecycle Management (pg. 133) as a part of the University of Michigan Executive Course on Product Lifecycle Management (PLM).

In this chapter, Sect. 1 has highlighted some of the significant importance of cryptography, E2E encryption, and RSA. Section 2 has highlighted different problems of RSA nowadays. Section 3 with its subsections has highlighted the following three things:

1. SHRSA-based [19–23] six-layered personal messaging communication protocol.
2. SHRSA's three-layer authentication and the problems which are solved by this multilayered authentication.
3. The problems are solved by SHRSA messaging scheme's communication protocol.

Section 4 has described the result analysis of the Secure Hybrid RSA (SHRSA)-based lightweight and efficient personal messaging communication protocol with special highlight to four distinguished features of SHRSA messaging communication protocol.

These four distinguished features are as follows:

1. Less memory usages of SHRSA messaging communication protocol than RSA.
2. Less CPU usages of SHRSA messaging communication protocol than RSA.
3. Efficiency in decryption than main RSA and CRT-RSA.

4. Security features provided by SHRSA messaging communication protocol.

Section 5 has concluded the chapter.

2 Current Scenarios

RSA algorithm has a very renowned property named multiplicative property. For example, if we consider for plaintext messages x_1 and x_2, as per this property, we can have encryption as shown in Eq. (1):

$$\mathrm{enc}_K(x_1 x_2) = \mathrm{enc}_K(x_1) * \mathrm{enc}_K(x_2) \tag{1}$$

Exploiting this homomorphic property [24] of RSA, Davida [1] has shown that main RSA is insecure to a chosen ciphertext attack. Judy Moore has shown us this in a simpler way, suppose an adversary is given a ciphertext $x = m^e \mod N$ and wants to compute m. Selecting a random $a \in Z_N$, the adversary asks for the plaintext of the ciphertext as shown in Eq. (2):

$$x_0 = xa^e \bmod N \tag{2}$$

Since $m_0 = x_0^d \bmod N = (xa^e)^d \bmod N = x^d a^{ed} \bmod N = ma \bmod N$

The adversary can very easily compute $m = m_0 a^{-1} \bmod N$ to recover the desired plaintext. This is a Chosen Ciphertext Attack (CCA).

Boneh et al. [7] have shown us another dangerous scientific problem called as exploitation of homomorphic property of RSA.

In a nutshell:

- Exploitation of multiplicative property and exploitation of homomorphic property (meet-in-the-middle attack) of RSA are very critical.
- Proper padding scheme [8], like Optimal Asymmetric Encryption Padding (OAEP) [16], is strong enough to battle the exploitation of homomorphic property and the exploitation of multiplicative property of RSA.

Integer factorization problem is a burning security issue for RSA. In 1998, Boneh and Venkatesan [5] have shown us with proper proof that the RSA tricky could be easier to crack, once the public exponent is small (or the product of small factors). Brown [25] in 2005 revealed that resolving the RSA tricky with a straight-line program is just about same tough like the integer factorization problem, on condition that the public exponent is small or has a small factor. With Pollard's general Number Field Sieve (NFS), following Lenstra [2, 26], we can use Eq. (3) as the heuristic anticipated runtime of the NFS to do the factorization of a composite number x.

$$L[x] = e^{1.923(\log x)^{1/3}(\log\log x)^{2/3}} \tag{3}$$

Again, with Lenstra [2, 26], we can find the heuristic anticipated runtime of the elliptic curve factorization method (ECM) with Eq. (4), to get a factor y of x.

$$E[x, y] = (\log_2 x)^2 e^{\sqrt{2(\log y)^{1/2} (\log \log y)^{1/2}}} \tag{4}$$

It is revealed that the anticipated runtime of the ECM is a function of the smallest factor along with the number which is being factorized. Here, the size of the smallest factor decides the complexity. We know that this complexity of the integer factorization tricky is dominating the security feature for RSA and the level of security of the RSA is resolute by the size of the modulus. So, nowadays:

- Factorization of the modulus with the NFS or the ECM is big bottleneck of RSA.
- In the present era, 1024-bit moduli are recommended for noncritical encryption as it necessitates, roughly 2^{80} operations, which is somehow very infeasible in the modern era.

Wiener has shown in [27] that picking up a private exponent to be too small is completely insecure. Wiener [3] showed in 1990 that any case of RSA with parameters satisfying Eq. (5) is absolutely breakable,

$$kdg < xy\left(\frac{2}{3(x + y)}\right) \tag{5}$$

where k is the constant in the following key Eq. (6) :

$$ed = 1 + k\lambda(N) \tag{6}$$

and $g = \gcd(x - 1, y - 1)$

Wiener [3] revealed that how we can factorize the modulus N for these instances very efficiently, by making use of the information acquired from the continuous fraction extension of e/N. In 1999, Boneh and Durfee [6] extended Coppersmith's lattice-based technique for finding small solutions of modular univariate polynomials [4, 9, 28] to modular bivariate polynomials, for the purpose of rising the bound on insecure private exponents to $d < N^{0.292}$. Blomer and May [28] in 2001 revealed a refined lattice-based attack which makes the analysis simpler, but without improvement of the Boneh and Durfee's bound [6, 18]. In 2004, Hinek [14, 27, 29–33] revealed that all of the attacks on small private exponent RSA, also workable with small negative private exponents, produce the same bounds. In 1996, the tricky of large private exponents (p > 1) was earliest well thought out by Chen, Chang, and Yang. Chen, Ku, and Yen [34] have revealed a new attack based on lattice, with the case of private exponents close to λ (N) in 2005. It was proved that these cases of RSA with private exponent d satisfying $|d - \lambda(N)| < N^{0.25}$ are unsafe, when the modulus is sufficiently large. In multi-prime RSA [16, 35], we take the modulus, $N = \prod_{i=1}^{a} x_i$, (the product of distinct primes where all primes are balanced primes) like if $x_i < x_i + 1$ for $i = 1, \ldots, a - 1$, then we can proceed as per Eq. (7) :

$$4 < (1/2N)^{1/a} < x_1 < N^{1/a} < x_a < (2N)^{1/a} \tag{7}$$

Here, we assume that the public and private exponents are well-defined modulo ϕ (N) as shown in Eq. (8) :

$$\phi(N) = \prod_{i=1}^{a}(x_i - 1) \tag{8}$$

Thus, e and d must satisfy Eq. (9):

$$ed \equiv 1(\text{mod}\,\phi(N)) \tag{9}$$

which we can take as the main relation. From this Eq. (9), we have the main Eq. (10):

$$ed = 1 + k\phi(N) \tag{10}$$

where k is some positive integer.

Generally, we can use $\wedge$ to signify the variance like $N = \phi(N) - \wedge$. From Eq. (8), we can get $\wedge$ as shown in Eq. (11):

$$\begin{aligned}\wedge &= N - \phi(N) = N - \prod_{i=1}^{a}(x_i - 1)\\ &= \sum_{i=1}^{a} N/x_i - \sum_{i,j=1,i<j}^{a} N/x_i x_j + \sum_{i,j,k=1,i<j<k}^{a} N/x_i x_j x_k + \ldots + (-1)^a\end{aligned} \tag{11}$$

We can see from [15, 36], a simple computation by use of Eq. (11) for $\wedge$ and 1.3 (condition for balanced primes) revealed that $\wedge$ fulfills $|\wedge| < (2a - 1)\, N^{1-1/a}$. Thus, ϕ (N) and N have roughly an (a − 1)/a fraction of their most significant bits in common. The encryption algorithm for multi-prime RSA is undistinguishable to that of RSA. The public (encrypting) exponent will commonly be symbolized by $e = N^{\alpha}$.

While the idea of performing decryption, using CRT [10–12, 36] was mentioned as early as 1977 in the RSA patent (when the modulus consists of more than two primes) and in 1979 by Rabin, it was not fully appreciated for use in RSA until J-J. Quisquater's and C. Couvreur's work in 1982 [10]. Divide-and-conquer technique gives CRT-RSA almost theoretical speedup up to four times faster than normal RSA as shown in Eq. (12):

$$S_{\text{CRT-RSA}} = \log^3 n/2(\log(n)/2)^3 = 4 \tag{12}$$

M-prime RSA just extends the CRT-RSA for the decryption like it calculates in Eq. (13):

$$M_i = C^{d_i} \bmod x_i \tag{13}$$

for $1 \leq i \leq k$. Then, by using CRT to Eq. (13), we can get the plain text $M = C^d \bmod n$. M-prime RSA has $d_i = O(n^{1/k})$ (so as to $\log d_i = O(\log(n)/k)$) and the multiplication cost of $O((\log(n)/k)^2)$ for a whole cost of $O(k(\log(n)/k)^3) = O(\log^3(n)/k^2)$, resulting in the theoretical speedup of M-prime RSA compared to CRT-RSA which is shown in Eq. (14):

$$S = k^2/4 \tag{14}$$

So,

- Computational modular exponentiation complexity problem and partial key exposure vulnerability can be very well tackled by M-Prime RSA and CRT-RSA combination.
- The asymptotic very low speed of decryption of RSA problem can be solved by M-Prime RSA and CRT-RSA combination.

Shared RSA [13] gives a good resolution of low modular complexity problem of RSA. As per shared RSA [13], in decryption, we can get enhanced complexity of RSA decryption.

Also:

- More effortlessness and speediness can come with efficient RSA [14] with extended Euclidean algorithm.
- In parallel, Perfect Forward Secrecy (PFS) using Diffie–Hellman and Diffie–Hellman key-exchange protocol can defend the keys in transit in peer-to-peer communications and can give parallel protection to sniffing attacks.

So, it is vivid that variants of RSA [1–18] along with some mathematical algorithms can solve some of the major issues of RSA in the scenario of personal messaging, for the existing Internet and future Internet architectures [37, 38].

3 Designing of the Secure Hybrid RSA (SHRSA)-Based Lightweight and Efficient Personal Messaging Communication Protocol

At first, we have taken some of the RSA variants and introduced the Secure Hybrid RSA (SHRSA)-based lightweight and efficient personal messaging communication protocol's encryption and decryption, with some other algorithms to resolve some of the major problems with main RSA, as shown in Fig. 1.

As an outcome, we have developed an SHRSA [19–23] messaging communication protocol. The most precarious scientific attacks on RSA is exploitation of multiplicative property as shown in [24]. We are using OEAP [24] added with some random

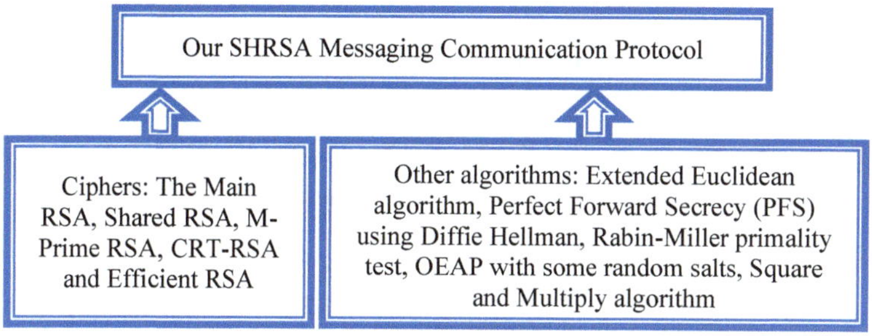

Fig. 1 The SHRSA messaging communication protocol's algorithms integration

salt with some synchronous time gap, with the plain text, before the SHRSA-based lightweight and efficient personal messaging communication protocol's encryption begins. This is resolving the below scientific attacks:

- The exploitation of multiplicative property of RSA (chosen ciphertext attack).
- Also, protecting us from exploitation of homomorphic property (meet-in-the-middle attack) as shown in [39].

As the expected complexity necessitates for the 1024-bit RSA modulus, approximately 2^{80} operations, which is very infeasible. So, we are also using 1024-bit RSA modulo. We are using square and multiply algorithm, in the SHRSA encryption, for decreasing the high cost of modular exponentiation computation. The messaging communication protocol is able to defend the Brute-force attack by changing the keys in synchronous time gap. Rabin–Miller Primality test is confirming us of strongest primes in the scheme.

Low modular complexity is resolved by incorporating shared RSA, in the SHRSA decryption, as shown in [13]. The SHRSA decryption is having more and more effortlessness for users, with high speed, by incorporating efficient RSA with extended Euclidean algorithm.

3.1 *SHRSA Messaging Communication Protocol*

The SHRSA-based six-layered personal messaging communication protocol is a multilayered protocol stack as shown in Fig. 2.

First layer end-to-end connection layer establishes pure dedicated end-to-end channel for each peer recognized by IP address. End-to-end connection layer is responsible for creating and receiving TCP connections from peers with TCP relay option, recognized by IP address used. Second layer is for providing multilayer authentication and This second layer shown in Fig. 2 is actually consisting of three layers of authentications internally. The three-layer authentications are shown in Fig. 3.

```
+------------------------------------------------------------+
| SHRSA Decryption Layer |
+------------------------------------------------------------+
| SHRSA Encryption Layer |
+------------------------------------------------------------+
| End to End (E2E) Secure Channel Layer|
+------------------------------------------------------------+
| 3-way Handshaking |
+------------------------------------------------------------+
| Authentication Layer |
+------------------------------------------------------------+
| End to End Connection Layer |
+------------------------------------------------------------+
```

Fig. 2 The SHRSA's six-layered communication protocol stack

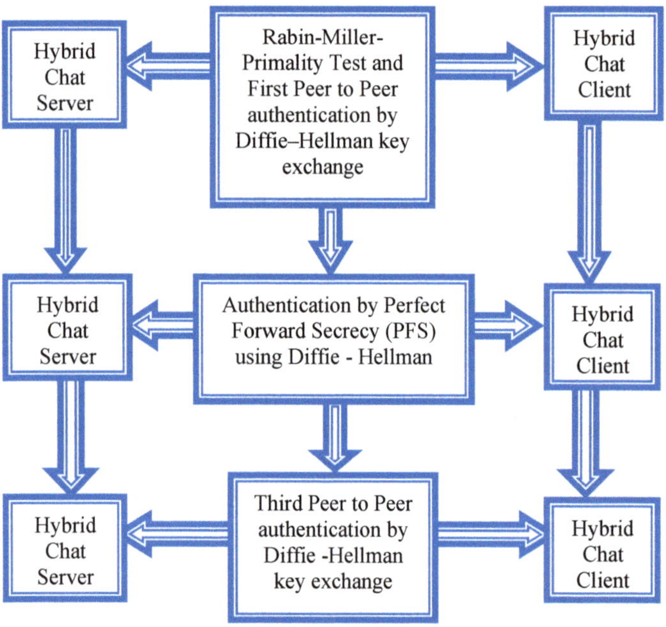

Fig. 3 The SHRSA's three-layered authentication

The first- and third-layer authentications in Fig. 3 are done by D–H key-exchange protocol [40] and second-layer authentication is done by PFS using Diffie–Hellman key exchange (D–H) as shown in Fig. 3. These approaches in the scheme enable us to protect the keys in transit in peer-to-peer communications and giving us parallel shielding to sniffing attacks.

Third layer in Fig. 2 is responsible for three-way handshaking between SHRSA server and SHRSA clients. The communication protocol works with n*n way (n number of SHRSA server and n number of SHRSA clients).

Now, layer 4 works for establishing E2E secure channel with only one peer (1 SHRSA server and 1 SHRSA client). In the secure communication protocol, multiple SHRSA servers and SHRSA clients can connect with each other but after connection it will stop in layer 3 in Fig. 2. Only one peer at time works with pure peer-to-peer way and goes to the layer 4 in Fig. 2.

3.2 Problems Resolved by SHRSA Messaging Communication Protocol

In the scheme's encryption (Layer 5 in Fig. 2), we have first integrated main RSA with Pohlig–Hellman encipher and with efficient RSA for more and more strong and statistical complexity. Before encryption starts, we are using OAEP [16] and random salts with the plaintext. This OAEP and random salts are helping us to resolve two major scientific problems of RSA. It is protecting us from the scientific attacks like exploitation of multiplicative property (chosen ciphertext attack), exploitation of homomorphism property (meet-in-the-middle attack), and short plaintext attack.

The scheme is having more and more effortlessness for users with high speed by use of efficient RSA with extended Euclidean algorithm. The encryption and decryption processes are identical to the original RSA, the computational rate due to encryption and decryption will not be different ominously from the original RSA.

Hence, the SHRSA encryption is also having the encryption complexity $= (3n_e - 2)(n^2 + 2)$.

In these ways, the scheme's encryption gives us following solutions:

1. Enable us to tackle dangerous known-plaintext attacks.
2. It is giving us more and more strong and statistical complexity.
3. It is giving us parallel protection to sniffing attacks and real-time key negotiation between each peer is given by PFS using D–H.
4. It is having more and more effortlessness for users with high speed by use of efficient RSA with extended Euclidean algorithm.
5. The computational rate due to encryption will not be different ominously from the original RSA. Therefore, the SHRSA encryption is also having the encryption complexity $(3n_e - 2)\ (n^2 + 2)$ alike main RSA.
6. It enables us to be protected from exploitation of multiplicative property and homomorphic property (meet-in-the-middle attack).
7. It is giving us resolution for difficulty of the integer factorization problem of RSA and very computationally costly exponentiation modulo N problem.
8. 1024-bit key is giving us solution for the exploitation of certain key choices problems.

9. RSA's high computational cost is totally irrelevant in all aspects of the Internet of Everything (IoE) scenarios. This high cost owing to modular exponentiation is getting decreased by the use of the square and multiply algorithm in the scheme's nine-layered protocol stacks.

Now, let us focus on the sixth layer of the messaging scheme's multilayered protocol stacks. SHRSA decryption layer is responsible for the SHRSA decryption of real-time messages in every chat session. Asymptotic very low decryption speed is the very big issue in present Internet scenario and for future Internet architectures also. So, we are inserting multi-prime RSA with CRT-RSA to address these issues. The divide-and-conquer technique boosts the CRT for high efficiency and multi-prime RSA again boosts up the decryption speed by decreasing the size of exponents and moduli. Again, we are inserting the shared RSA in the SHRSA decryption to have more complexity to make it more secure, strong, and reliable. Hence, the SHRSA cipher's decryption algorithm is resolving the issues like:

- Low modular complexity issue of RSA.
- Computational modular exponentiation complexity and partial key exposure vulnerability.
- Asymptotic very low speed of decryption of RSA (multi-prime RSA and CRT-RSA insertion are resolving these scientific problems). In the SHRSA decryption algorithm, we are getting almost nine times faster decryption than normal RSA, with a proper balance of security, privacy, and efficiency.

In these ways, the SHRSA-based six-layered personal messaging communication protocol is highly efficient, reliable, and strong, with balance of the major issues like security, privacy, and authentication with high efficiency.

4 Result Analysis of Secure Hybrid RSA-Based Six-Layered Personal Messaging Communication Protocol

The SHRSA messaging scheme's multilayered secure communication protocol is designed in Java and we are using NetBeans IDE 8.0.2. For SHRSA encryption and SHRSA decryption, we have 21 classes for each one as whole 42 classes. In the software package AB_HYBRID_RSA_MESSEGING_SCHEME_SERVER_FOR_FUTURE_INTERNET_ARCHITECTURES, we have SHRSA cipher chat server and client, which are responsible for secure hybrid RSA encryption and decryption. Prime generator class takes care of Rabin–Miller Primality test for strongest, secure and safest, and larger primes. Diffie–Hellman initiator class and challenger class with respondent class take care about PFS based on DH, peer-to-peer authentication by D–H key exchange, three-way handshaking, and D–H key exchange as shown in Fig. 2. Moreover, these three classes are responsible for three-layered authentications as

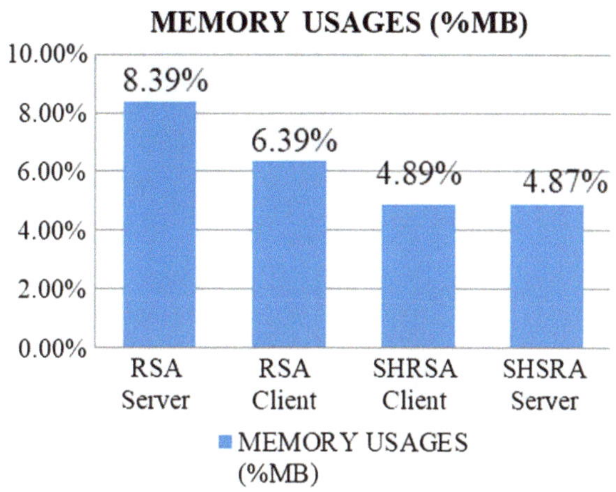

Fig. 4 RSA with SHRSA client and SHRSA server's memory usage comparisons

shown in Fig. 3. Furthermore, these three classes form pure mesh topology to ensure pure peer-to-peer and E2E authentication with the ciphers class. Along with the SHRSA CipherChatClient and SHRSACipherChatServer classes and left all connected classes work for SHRSA messaging encryption layer and SHRSA messaging decryption layer. These classes are the main classes for secure encrypted messaging and decryption of those messages. In the past works [20, 21], we have already tested with n number of SHRSA clients and n number of SHRSA servers trying to communicate with each other as peers. But we have shown that at a time only one peer can start secure end-to-end authenticated messaging and other all peers will be connected to each other but must wait for present communicating peers to end first. We have first checked the memory performance of the SHRSA messaging scheme's SHRSA server and client with RSA server and client with the setup desktop with Intel(R) Core(TM) i7-4790 K CPU @ 4.00 GHz (8 CPUs), ~4.0 GHz. For testing comparison purpose, we have developed the API for RSA server and client messaging. Already, we have the APIs for SHRSA server and client with multilayered secure communication protocol stacks. Now, here we have discussed the protocol's memory occupancy testing for main RSA server and client messaging API and SHRSA client and server API. The allocated memory for testing purpose was 512 MB. Now, we have run all these four APIs and did constant messaging vice versa for 2 minutes. The memory occupancy data (in %) what we got is shown in Fig. 4. Here, we have found that the SHRSA's multilayered secure communication protocol's server and client APIs at the time of messaging are very less memory occupier than the main RSA cipher. Detailed result calculations are in Appendix A.

Then, during this 2 min continuous messaging testing period, we have tested the CPU usages of the SHRSA server and client and main RSA server API and client API. We have found that again the SHRSA client and server are much less CPU occupier than main RSA instant messaging server and client. The CPU occupancy data (in %) what we got is shown in below Fig. 5. Here, we have found

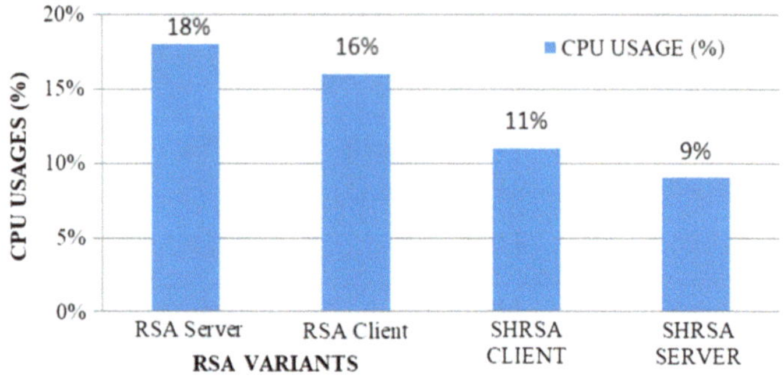

Fig. 5 RSA with SHRSA client's and SHRSA server's CPU usages comparisons

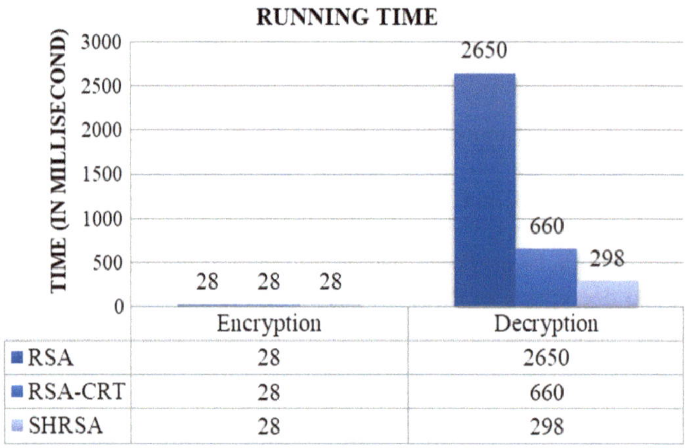

Fig. 6 Running time comparisons of the protocol and main RSA and CRT-RSA

that the SHRSA cipher as a whole is very less CPU occupier than the main RSA cipher. The protocol works with N = 1024 bit and text block size is from 1 < text block size < 255 bytes.

At the testing time, we have found the running time for encryption and decryption of each cipher's protocol (RSA, CRT-RSA, SHRSA) as shown in Fig. 6. These encryption and decryption times are in millisecond and the result times shown in Fig. 6 are the average times of 10 times running of encryption and decryption of each of API of RSA, CRT-RSA, and SHSRSA. So, it is clear that the SHRSA decryption running time is so less than CRT-RSA and main RSA.

So, these less memory occupancies, less CPU occupancies, then gaining in decryption make the scheme very relevant to these days Internet and IoE's need.

Now, if we compare the SHRSA cipher's protocol's encryption complexity and decryption complexity with main RSA and CRT-RSA, it can be shown in Table 1. Though RSA and CRT-RSA decryption complexity calculations are known to us, still we have calculated again in Appendix B. Details of calculation of SHRSA cipher's

Table 1 Comparisons of SHRSA cipher's encryption complexity and decryption complexity with main RSA and CRT-RSA

Comparison	RSA	CRT-RSA	SHRSA
Encryption complexity	$(3n_e - 2)(n^2 + n)$	$(3n_e - 2)(n^2 + n)$	$(3n_e - 2)(n^2 + n)$
Decryption complexity	$3n^3 + o(n^2)$	$\frac{3n^3}{4} + o(n^2)$	$\frac{n^3}{3} + o(n^2)$

Table 2 Comparisons of the complexity of decryptions of variants of RSA

Variant	Complexity order	Theoretical gain for n = 1024 bit, k = 3	Unit time for decryption
RSA without CRT	$O(n^3)$	1.0	1
RSA with CRT	$2 \cdot O(n/2)^3)$	4.0	1/4 = 0.25
SHRSA decryption	$k.O\,((n/k)^3)$	$k^2 = 9$	1/9 = 0.11

encryption complexity and decryption complexity are in Appendix C. Detailed calculations of Table 1 are in Appendix B and Appendix C.

For 1024-bit SHRSA, k = 3 is the case we have used, which offers huge gain in decryption comparison to the standard RSA and CRT-RSA decryption. The RSA, CRT-RSA, and the SHRSA decryption complexities and gain calculations are in Appendix B, and summarized values are shown in Table 2 (n = 1024, k = 3). Here, we can see that the SHRSA's communication protocol's decryption gain is nine times than the RSA. Detailed calculations of Table 2 are in Appendix B.

Now, let us discuss some security features of the SHRSA cipher and how it can protect us from several dangerous attacks.

(1) Factorization of RSA modulus n

We are using Monte Carlo factor with Pollard p-1 factorization for resolving the factorization of RSA modulus n problem of main RSA. The prime number is decided by the own SHRSA big integer class with Rabin–Miller Primality test. Moreover, the salting process is enhancing the ability for protection from factorization of RSA modulus n problem.

(2) Chosen plaintext attack

We are using a padding scheme—OAEP added with some random salt with some synchronous time gap for resolving this attack and resolving the exploitation of multiplicative property of RSA (chosen ciphertext attack) problem. Likewise, this is helping us to enhance the scheme's security and privacy, by protecting us from exploitation of homomorphic property (meet-in-the-middle attack) on RSA. Apart from these, also we are using another layer of padding by using PKCS#5 for extra protection from this attack. Additionally, we are using one-way function to stop the chosen ciphertext attack.

(3) Broadcast decryption by low exponent attack

To avoid this attack, each pair of public keys $P_i = (e_i, n_i)$ and $P_j = (e_j, n_j$ and any broadcast message should satisfy $e_i \neq e_j$ or $m^{e_i}, m^{e_j} > n_i n_j$. The m and n are large enough, so product of m*n is again large, so in this way, we are avoiding this attack. For another similar attack—Broadcast Decryption by Common Modulus Attack, the message senders are not sending identical messages to receivers with the same modulus and relatively prime encryption exponents.

(4) Fault injection attack

This attack is an attack where the fault in the computation of the private key leads to the computation of the private key. As we know, in CRT-RSA, the signature generation is made of two exponentiations $S_p = m^{d_p} \bmod p$ and $S_q = m^{d_q} \bmod q$ (here $d_p = d \bmod p - 1$ and $d_q = d \bmod q - 1$). The signature is then acquired by use of the Garner's formula:

$$S = S_q + q\left(i_q\left(S_p - S_q\right) \bmod p\right)$$

where $i_q = q^{-1} \bmod p$.

Now, we can presuppose that a fault is injected for the duration of the computation of S_p resulting in a faulty signature $\tilde{S}$. Since $S \equiv S_p \bmod p$ and $S \equiv S_q \bmod q$. Here, dangerous thing is that the faulty signature $\tilde{S}$ will satisfy $\tilde{S} \equiv S \bmod q$ and $\tilde{S} \equiv S \bmod p$. So, very dangerously, the secret parameter q can be straightforwardly calculated by computing the gcd of S - $\tilde{S}$ mod and N. Thereafter, the private key can be obtained after finding q. We are resolving this attack by checking the correctness of the signature before outputting it. More precisely, we are checking the signature is returned if only if $S^e \bmod n = m$.

(5) Small difference between primes p and q attack the primes are already maintaining the logic of large difference to avoid this attack. The SHRSA big integer class is taking care of this.
(6) Finding the eth root attack

We are using the CRT to decrypt. Finding the eth root is a difficult problem [4, 33, 34], if n is large. But if if $\varphi(n)$ is given, it can be found in polynomial time. We are tackling this problem with $e = 3$ with CRT.

(7) Common prime attack

We are using separate SHRSA prime number generator class for getting strongest, safest, and larger prime to avoid this attack.

(8) Exploitation of multiplicative property problem of RSA

We are using padding scheme the RSA-OAEP added with some random salt with some synchronous time gap for getting rid of this attack.

(9) Exploitation of homomorphic property

We are using padding scheme the RSA-OAEP added with some random salt with some synchronous time gap for getting rid of this attack.

(10) Integer factorization problem and problem of large private exponents

We are tackling these attacks with 1024-bit RSA-based SHRSA big integer class and we are having the customized Monte Carlo factor with Pollard p-1 factorization for this and the prime number is decided by the SHRSA big integer class with Rabin–Miller Primality test.

(11) RSA exponent and efficiency problem and RSA private exponent problem

We are resolving this problem by using the SHRSA big integer class and we are having the customized Monte Carlo factor with Pollard p-1 factorization for this and the prime number is decided by the SHRSA big integer class with Rabin–Miller Primality test and of course we are using M-prime RSA with CRT-RSA for faster decryption. Moreover, we are using PKCS #5 for defining the exponents as inverses λ (N) = LCM (x − 1, y − 1).

Also, OEAP insertion with random salts in the multilayered SHRSA cipher is helping us to stop attacks such as Algebraic attacks, Hastad attack, Desmedt–Odlyzko attack, related message attacks, fixed pattern RSA signature forgery, and two attacks by Bleichenbacher.

Moreover, the scheme is Set Partial Domain One. Therefore, these less memory occupancies, less CPU occupancies, and then increasing the computational speed at decryption with faster running time make the SHRSA communication protocol very relevant to these days Internet and IoE's need.

Therefore, in these ways, we have achieved the high level of security using SHRSA cipher also.

Using PFS, the protocol is able to protect the messages from non-repudiation attack, man-in-the-middle attack, chosen cipher attack, and replay attack.

Hence, at the encryption level, the SHRSA encryption with 1024-bit RSA modulus, is helping us to resolve some of the scientific problems like the exploitation of multiplicative property, the exploitation of homomorphic property (meet-in-the-middle attack), difficulty of the integer factorization problem of RSA, the very high computationally costly exponentiation modulo N problem, and low modular complexity with effortlessness and speediness problem. Also, the SHRSA encryption scheme has proper protection from chosen plaintext attack and short plaintext attack, etc., along with protection to sniffing attacks and resolving the real-time key negotiation issue also. Brute-force attack is countered by randomly altering the keys in synchronous time slot with 1024-bit value.

In the decryption level of SHRSA, the SHRSA decryption is helping us to resolve some of the scientific problems like computational modular exponentiation complexity, partial key exposure vulnerability, and asymptotic very low speed of decryption of RSA problem.

5 Conclusions

The SHRSA-based lightweight and efficient personal messaging communication protocol is a perfect combination of strong security, authentication, and reliability. It can protect the private information with a strong balance of efficiency, with a blend of the messaging communication protocol's encryption and decryption. The scheme can resolve some of the important scientific problems of RSA, for using it in personal messaging scenarios in present Internet and future Internet architectures. In the encryption level, the SHRSA messaging scheme's encryption algorithm with 1024-bit RSA modulus is helping us to resolve some of the scientific problems like the exploitation of multiplicative property, the exploitation of homomorphic property (meet-in-the-middle attack), difficulty of the integer factorization problem of RSA, the very high computationally costly exponentiation modulo N problem, and low modular complexity with effortlessness and speediness problem. Moreover, the SHRSA messaging communication protocol's encryption has proper protection from chosen plaintext attack and short plaintext attack, etc., along with protection to sniffing attacks and resolving the real-time key negotiation issue also. Brute-force attack is countered by randomly altering the keys in synchronous time slot with 1024-bit value. In the decryption level, the SHRSA messaging communication protocol's decryption is helping us to resolve some of the scientific problems like computational modular exponentiation complexity, partial key exposure vulnerability, and asymptotic very low speed of decryption of RSA problem. We are gaining nine times asymptotic decryption speed than the RSA. We have also seen in the testing that the communication protocol is occupying very less memory and it is less CPU occupier also than RSA and CRT-RSA. These properties of the scheme make it more accepted in low-memory and low-CPU constraint environs. All these properties we have seen in results have made the protocol very relevant to all aspects of the IoE scenarios.

Acknowledgements This research is supported by National Natural Science Foundation of China (No. 61631013). We want to convey our gratitude and tribute to Late Prof. Wang Jing for his constant supervision and encouragement for this project.

Appendix A: RSA with SHRSA Client and SHRSA Server's Memory Usage Comparisons

We are getting % of memory usage like this, for example, SHRSA client memory usage was 24.80 MB out of 512 MB available during the testing. So, usage percentage is (24.93/512)*100 = 4.8691%

RSA variants	Memory usage percentage (%)	Available memory (MB)	Occupied memory (MB)
RSA server	8.39	512	42.96
RSA client	6.39	512	32.72
SHRSA client	4.89	512	25.03
SHSRA server	4.87	512	24.93

Appendix B: Decryption Complexities Calculations

RSA

Decryption method:

Following parameters are used:

- n = number of bits in modulus.
- n_e = number of bits in public exponent (e).
- n_d = number of bits in private exponent (d).

Decryption method:

1. Uses his private key (N, d) to compute

   $M = C^d \bmod N.$

2. Extracts the plaintext from the message representative M.

   Here, the iteration is done n_d times (n_d = No of bits in d)
   and $n_d \approx n$, So:
   Decryption complexity $= (3n - 2)(n^2 + n) \approx 3n^3 + o(n^2)$

CRT-RSA

Decryption method:

1. Calculate $d_p = d \bmod p\text{-}1$ and $d_q = d \bmod q\text{-}1$.
2. Calculate $M_p = C^{d_p} \bmod p$

   and $M_p = C^{d_q} \bmod q$.

3. Calculate M from M_p and M_q using CRT.

   Complexity of decryption algorithm $\approx 3n^3/4 + o(n^2)$

SHRSA

Decryption method:

The RSA modulus was modified so that it can further decrease the decryption time. It consists of k primes $p_1, p_2, \ldots, p_k$ instead of using only two.

For computation, we have considered k = 3,

1. Calculate $d_p = d \bmod p\text{-}1$, $d_q = d \bmod q\text{-}1$, and $d_r = d \bmod r\text{-}1$.
2. Calculate $M_p = C^{d_p} \bmod p$, $M_q = C^{d_q} \bmod q$, $M_r = C^{d_r} \bmod r$.
3. Calculate M from M_p, M_q, and M_r using CRT

SHRSA decryption $= k^*(n_d+1/2n_d)n_p^2 = k*(n/k + n/2k)(n/k)^2 = 1/k^2(2/3n^3)$

So far, we have considered k = 3, so we are getting approximately 2.25 gain than CRT-RSA.

And compared to main RSA, we are gaining = 9 times(as CRT-RSA is gaining 4 times g than RSA, so SHRSA gains 4*2.25 = 9)

$$\text{Decryption complexity}_{(SHRSA)} = (3 * (n - n/3)(n/3 + 2)) + \left(3 * \left(3 * (n/3)^3 + (n/3)^2\right) + 16^{n^2/3} + o\left(n^2\right)\right. \approx n^3/3 + o(n^2)$$

Some more on this SHRSA has attained a decryption speedup compared with plain RSA and CRT-RSA, just by decreasing the size of exponents and moduli, at the cost of extra modular exponentiations. Though a linear rise in the number of exponentiations turns to a cubic reduction in the cost of each exponentiation for a complete speedup that is quadratic in the number of factors k of the modulus. Properly evaluating $C^d \bmod n$ for d = O(n) costs $O(\log^3 n)$, while SHRSA has $d_i = O(n^{1/k})$ (so that $\log d_i = O(\log(n)/k)$) and multiplication cost of $O((\log(n)/k)^2)$ for a complete cost of $O(k(\log(n)/k)^3) = O(\log^3(n)/k^2)$. We have used three prime numbers as the scheme's modulus size (bits) is 1024 bits.

Ever since CRT-RSA is 4 times as fast as main RSA by this time, the theoretical speedup of the scheme's decryption comparison with RSA is as shown below:

$\boxed{D_{SHRSA} = k^2/4 = 9/4 = 2.25}$ (as we have considered k = 3 as mentioned in Appendix B earlier).

Standard RSA decryption using CRT necessitates two complete exponentiations modulo n/2 bit numbers. In the scheme's decryption, it necessitates b full exponentiations modulo n/b bit numbers. In case of d is on the order of p, the execution time is $O(\log^3 x)$. As a result, the asymptotic speedup of the SHRSA decryption comparison to the standard RSA is as shown below again:

$$2.(n/2)^3/b.(n/b)^3 = b^2/4$$

One of the advantages of SHRSA decryption is time, by use of the Chinese Remainder Theorem and doing the calculations in parallel, the number of bit operations necessitates to decrypt a ciphertext is at most

$$\frac{3}{2r^3}\left(\log_2 N\right)^3 \text{ (Using standard arithmetic)}$$

So, the time necessary for decryption gets lower with each additional prime in the modulus. Another advantage of the scheme is space, by use of the Chinese Remainder Theorem, the space needed for all decryption computations until the very last (recombining step) require only $\left(\log_2 N\right)/r$ space, where p_r is the largest prime in the modulus. If all the primes are roughly $(\log_2 N)/r$-bits large (balanced primes), the space required decreases with each additional prime added to the modulus.

Appendix C: SHRSA Encryption and Decryption Complexities Calculations

The RSA modulus was modified so that it can further decrease the decryption time. It consists of k primes $p_1, p_2, \ldots, p_k$ instead of using only two.

For computation, we have considered $k = 3$,

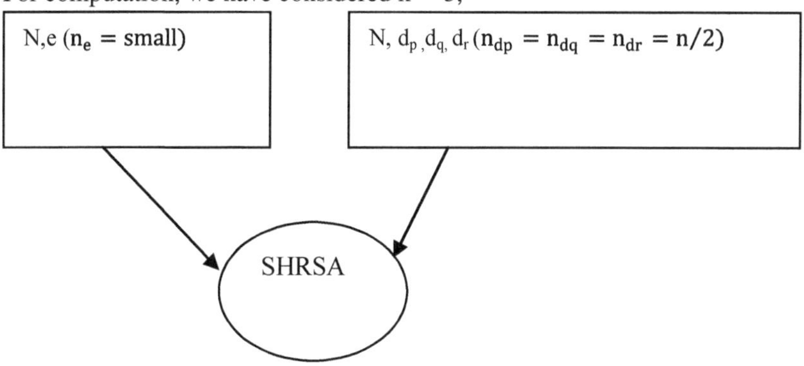

Following parameters are used:

- n = Number of bits in modulus.
- n_e = Number of bits in public exponent (e).
- n_d = Number of bits in private exponent (d).

Key generation method: k = Number of primes to be used.

1. Compute k distinct primes $p_1, \ldots, p_k$ each one [log N/k] bits in length and $N = \pi_{i-1}^{k} p_i$.
2. Compute e and d such that $d = e^{-1} \bmod \varphi(N)$, where $\gcd(e, \varphi(N)) = 1$, $\varphi(N) = \pi_{i-1}^{k}(p_i - 1)$
3. For $1 \le i \le k$, compute $d_i = d \bmod (p_i\text{-}1)$.

Public key = (N, e).
Private key = $(d_1, d_2, \ldots, d_k)$.

Decryption method:

1. Calculate $d_p = d \bmod p\text{-}1$, $d_q = d \bmod q\text{-}1$, and $d_r = d \bmod r\text{-}1$.
2. Calculate $M_p = C^{d_p} \bmod p$, $M_q = C^{d_q} \bmod q$, and $M_r = C^{d_r} \bmod r$.
3. Calculate M from M_p, M_q, and M_r using CRT

$$\begin{aligned}\text{Decryptioncomplexity}_{(\text{SHRSA})} &= (3*(n-n/3)(n/3+2)) \\ &+ \left(3*\left(3*(n/3)^3+(n/3)^2\right)\right. \\ &+ 16^{n^2/3}+o\left(n^2\right) \\ &\approx n^3/3+o(n^2).\end{aligned}$$

References

1. Davida, G.I.: Chosen signature cryptanalysis of the RSA (MIT) public key cryptosystem. Technical Report TR–CS–82–2, Department of EECS, University of Wisconsin, Milwaukee, October 1982. http://www.uwm.edu/~davida/papers/chosen/
2. Lenstra: Factoring integers with elliptic curves. Ann. Math. **126**, 649–673 (1987)
3. Wiener, M.J.: Cryptanalysis of short RSA secret exponents. IEEE Trans. Inf. Theory **36**(3), 553–559 (1990)
4. Coppersmith, D.: Finding a small root of a univariate modular equation. In: Advances in Cryptology—EUROCRYPT '96, volume 1070 of Lecture Notes in Computer Science, pp. 155–165. Springer (1996)
5. Boneh, D., Venkatesan, R.: Breaking RSA may not be equivalent to factoring. In: Advances in Cryptology - EUROCRYPT '98, volume 1403 of Lecture Notes in Computer Science, pp. 59–71. Springer, 1099
6. Boneh, D., Durfee, G.: Cryptoanalysis of RSA with private key d less than n 0.292. IEEE Trans. Info. Th. **46**(4), 1339–1349 (2000)
7. Boneh, D., Joux, A., Nguyen, P.Q.: Why textbook ElGamal and RSA encryption are insecure. In: Advances in Cryptology—ASIACRYPT 2000, volume 1976 of Lecture Notes in Computer Science, pp. 30–44. Springer, (2000). Boneh, D., Shacham, H.: Fast variants of RSA. In: CryptoBytes **1**(5), 1–9 (2002)
8. Boneh, D., Shacham, H.: Fast variants of RSA. CryptoBytes **1**(5), 1–9 (2002)
9. Coron, J.S.: Finding small roots of bivariate integer polynomial equations revisited. In: Advances in Cryptology—EUROCRYPT 2004, volume 3027 of Lecture Notes in Computer Science, pp. 492–505. Springer (2004)
10. Wagner: Cryptanalysis of a provably secure CRT-RSA algorithm. In: Proceedings of the 11th ACM Conference on Computer and Communications Security, pp. 92–97 (2004)
11. Sun, H.M., Wu, M.E.: Design of rebalanced RSA-CRT for fast encryption. In: Information Security Conference (2005)
12. Nguyen, H.L.: RSA threshold cryptography. In Technical Report, Department of Computer Science, University of Bristol (2005)
13. Hinek, M.J., Stinson, D.R.: An inequality about factors of multivariate polynomials. In CACR Technical Report CACR 2006–15, Centre for Applied Cryptographic Research, University of Waterloo (2006)
14. Alhasib, A., Haque, A.L.: A comparative study of the performance issues of the AES and RSA cryptography. In: Proceedings 3rd International Conference on Convergence and Hybrid Information Technology (ICCIT), Busan, pp. 505–510 (2008)
15. Oleshchuk, V.: Internet of things and privacy preserving technologies. In: Wireless Communication, Vehicular Technology, Information Theory and Aerospace& Electronics Systems Technology, Aalbrg, pp. 336–340 (2009)
16. Ma, K., Liang, H., Wu, K.: Homomorphism property-based concurrent error detection of RSA: a countermeasure to fault attack. IEEE Trans. Comput. **61**(7), July 2012
17. Shelby, Z., Hartke, K., Bormann, C.: The constrained application protocol (coap). RFC 7252, Internet Engineering Task Force, Jun. 2014
18. Xiao, Z., Wang, Y., Jiang, Z.: Research and implementation of four-prime RSA digital signature algorithm. In: 2015 IEEE ICIS 2015, June 28-July 1 2015, Las Vegas, USA

19. Bhattacharjya, A., Zhong, X., Wang, J., et. al.: On Mapping of Address and Port using Translation (MAP-T). Int. J. Inf. Comput. Secur. **11**(3), 214–232 (2019)
20. Bhattacharjya A., Zhong X., Wang J.: Hybrid RSA based highly efficient, reliable and strong personal Full Mesh Networked messaging scheme. Int. J. Inf. Comput. Secur. **10**(4), 418–436 (2018)
21. Bhattacharjya A, Zhong X, Wang J, et al.: Secure IoT Structural Design for Smart Homes, Smart Cities Cybersecurity and Privacy. Elsevier. pp. 187–201. (2019) http://www.sciencedirect.com/science/article/pii/B9780128150320000135..
22. Bhattacharjya A, Zhong X, Wang J, et al.: Security Challenges and Concerns of Internet of Things (IoT), Cyber-Physical Systems: Architecture, Security and Application. EAI/Springer Innovations in Communication and Computing, 153–185 (2019)
23. Bhattacharjya, A., Zhong, X., Wang, J., Xing, L.: CoAP—application layer connection-less lightweight protocol for the Internet of Things (IoT) and CoAP -IPSEC Security with DTLS Supporting CoAP. Accepted chapter in Book entitled "Digital Twin Technologies and Smart Cities" - Springer Series Title: Internet of Things (IoT)". CiteScore 0.88, IDS Number: BK0ZF
24. Bradly, J., Barbier, J., Handler, D.: Embracing the Internet of Everything to Capture Your Share of $ 14.4 Trillion. White Paper, Cisco (2013)
25. Hinek, M.J.: Small private exponent partial key-exposure attacks on multi-prime RSA. In Technical report, Citeseer (2005)
26. Lenstra, H., Lenstra, W., Lov´asz, L.: Factoring polynomials with rational coefficients. Mathematische Annalen **261**, 515–534 (1982)
27. Hinek, M.J.: Lattice attacks in cryptography: a partial overview. In: CACR Technical Report CACR 2004-08, Centre for Applied Cryptographic Research, University of Waterloo (2004)
28. Brown, D.R.L.: Breaking RSA may be as difficult as factoring. Cryptology ePrint Archive, Report 2005/380 (2005)
29. Hinek, M.J.: New partial key exposure attacks on RSA revisited. In: CACR Technical Report CACR 2004–02, Centre for Applied Cryptographic Research, University of Waterloo (2004)
30. Hinek, M.J.: (Very) large RSA private exponent vulnerabilities. In: CACR Technical Report CACR 2004-01, Centre for Applied Cryptographic Research, University of Waterloo (2004). Chen, C.Y., Ku, C.Y., Yen, D.C.: Cryptanalysis of large RSA exponent by using the LLL algorithm. Appl. Math. Comput. **169**, 516–525 (2005)
31. Sun, H.M., Wu, M.E. 'An approach towards rebalanced RSACRT with short public exponent'. In Cryptology ePrint Archive, Report 2005/053, 2005
32. Hinek, M.J.: Another look at small RSA exponents. In Topics in Cryptology—CT-RSA 2006, volume 3860 of Lecture Notes in Computer Science, pp. 82–98. Springer (2006)
33. Jochemsz, E., May, A.: A strategy for finding roots of multivariate polynomials with new applications in attacking RSA variants. In: Advances in Cryptology—ASIACRYPT 2006, volume 4284 of Lecture Notes in Computer Science, pp. 267–282. Springer (2006)
34. Blomer, May, A.: A tool kit for finding small roots of bivariate polynomials over the integers. In: Advances in Cryptology—EUROCRYPT 2005, volume 3494 of Lecture Notes in Computer Science, pp. 251–267 Springer (2005)
35. Bhattacharjya, A., Zhong, X., Wang, J.: Strong, efficient and reliable personal messaging peer to peer architecture based on Hybrid RSA. In: Proceedings of the International Conference on Internet of Things and Cloud Computing (ICC 2016) ISBN 978-1-4503-4063-2/16/03. The Møller Centre-Churchill College, Cambridge (2016). https://doi.org/10.1145/2896387.2896431
36. Turner, C.S.: Euler's Totient function and public key cryptography. Nov 7, 2008. Leusse, D., Periorellis, P., Dimitrakos, P.: Self-managed security cell a security model for the future internet architectures and services advances in future internet. In: Proceedings First International Conference on Digital Object Identifier, pp. 47–52 (2009)
37. Medaglia, C.M., Serbana, T.: An overview of privacy and security issues in the internet of things. In: II The Internet of Things'. In 20th Tyrrhenian Workshop on Digital Communications, New York: Springer New York, 2010 389–394

38. Braun, B.M.: Crowcroft, J. SNA: Sourceless Network Architecture, Technical Report, Number 849, Computer Laboratory, UCAM-CL-TR-849, ISSN 1476-2986, March 2014
39. Paxson, V., Sommer, R.: An architecture for exploiting multi-core processors to parallelize network intrusion prevention. In: Proceedings of the IEEE Sarnoff Symposium, pp. 1–7 (2007)
40. Diffie, W., Hellman, M.: "New directions in cryptography" (PDF). IEEE Trans. Inf. Theory **22**(6), 644–654 (1976)

Aniruddha Bhattacharjya is with the Department of Electronic Engineering, Tsinghua University, Beijing, China, as a Chinese Government Ph.D. scholar. His research interests are cryptography, network security, RFID-based architectures and middleware, security in fixed and wireless networks, applications of cryptography, and IoT security. He has received the ICDCN 2010, Ph.D. Forum Fellowship. He achieved the best paper award in ACM ICC 2016, in Cambridge University, UK. Since 2012, he has been working as an IEEE mentor and ACM faculty sponsor. He is a member of 34 IEEE societies and various IEEE technical committees. He has published 35 papers as well as 1 Chinese innovation patent is pending.

Xiaofeng Zhong received his Ph.D. in Information and Communication Systems from Tsinghua University in 2005. He is an Associate Professor in the Department of Electronic Engineering at Tsinghua University. He performs research in the field of mobile networks, including users' behaviors and traffic model analyses, MAC and network protocol design, and resource management optimization. He has published more than 30 papers and holds 7 patents.

Jing Wang received his BS and MS degree in Electronic Engineering from Tsinghua University, Beijing, China in 1983 and 1986, respectively. He has worked as a Faculty Member in Tsinghua University since 1986. He is currently a Professor at the School of Information Science and Technology. He also serves as the Vice Director of the Tsinghua National Laboratory for Information Science and Technology. His research interests are in the area of wireless communications, including transmission and networking technologies of 4G/5G. He has published more than 150 conference and journal papers.

Xing Li received the B.S. degree in radio electronics from Tsinghua University, Beijing, China, in 1982, and the M.S. and Ph.D. degrees in electrical engineering from Drexel University, Philadelphia, PA, USA, in 1985 and 1989, respectively. He is currently a Professor with the Electronic Engineering Department, Tsinghua University. His research activities and interests include statistical signal processing, multimedia communication, and computer networks. He has published more than 300 papers in his research areas. He is a Deputy Director of the China Education and Research Network (CERNET) Center and a Member of the Technical Board of the CERNET Project. He was a Member of Communication Expert Committee of the China National "863" High Technology Project. He is a Formal Chairman of the Asia Pacific Networking Group and a Formal Member of the executive council of the Asia Pacific Network Information Center.

MIX
Papier aus verantwortungsvollen Quellen
Paper from responsible sources
FSC® C105338

If you have any concerns about our products,
you can contact us on
ProductSafety@springernature.com

In case Publisher is established outside the EU,
the EU authorized representative is:
Springer Nature Customer Service Center GmbH
Europaplatz 3, 69115 Heidelberg, Germany

Printed by Libri Plureos GmbH
in Hamburg, Germany